KU-030-047

genes

THE FIGHT FOR LIFE

Z4652
Genes: The Fight for Life.

R.C.S. GUILDFORD,
J. C. MALLISON
LIBRARY.

genes

THE FIGHT FOR LIFE

BRIAN J. FORD

ACC No	CLASS No
24652	576.5
12/99	

CASSELL

Cassell
Wellington House
125 Strand
London WC2R 0BB
www.cassell.co.uk

Copyright © Brian J. Ford 1999
www.sciences.demon.co.uk

All rights reserved. No part of this publication may be reproduced or transmitted, in any
form or by any means, electronic, mechanical, including photocopying, recording or any
information storage and retrieval system, without the prior permission in writing from
the publishers and copyright owner.

The moral right of the author has been asserted

Distributed in the United States by
Sterling Publishing Co. Inc.
387 Park Avenue South
New York NY 10016–8810

British Library Cataloguing-in-Publication Data
A catalogue record for this book is available from the British Library

ISBN 0-304-35019-2

Designed by Geoff Green
Typeset by Geoff Green Book Design, Cambridge
Printed and bound in Great Britain by MPG Books Ltd
Bodmin, Cornwall

BRIAN J. FORD'S admirable new book should be come required reading for anybody who cares about the powers that humankind can now exercise to modify and understand its own nature. I read a draft while travelling on public transport and I consider it a real tribute to the subject matter that, caught up in the kaleidoscope of ideas, I twice went past my intended destination. However, the enforced stop-go of my reading helped me to digest the wealth of material and anecdote that is presented. I also share with the author the antipathy to the take-over that molecular biology has staged within the biological sciences, at the expense of understanding how whole creatures work. He points out that our bodies are made of a collection of living cells, which in their behaviour continue to practise many of the skills that they acquired at a very much earlier stage of their evolution. Most fascinating is the transference of the principles of immunology: the recognition of non-self at a cellular level, to the behaviour of societies, where tribalism rejects the stranger, often with devastating effect. All good stuff.

Professor Heinz Wolff

Contents

Acknowledgements

T o   l i s t   every person whose wise counsel and influence have bettered this book would be impossible. It would be equally difficult to acknowledge all the universities and institutes I have visited around the world, though I owe a special debt to the staff at the universities of Cambridge and Cardiff. Just for once, I would like to confine myself to mentioning some of those who have advised and assisted with processing this book: my colleague Professor Heinz Wolff DSc for digesting the first draft; Roger Pfister BSc for his continuing kindly advice as a computer specialist; my daughter Tamsin Ford BSc for working on the page proofs and three-year-old Lucy for her picture on p. 229; David Stone for his graphic processing skills; and Irene Williams for acting as courier. I am grateful to the editors of *The Times, Nature,* the *Mensa Magazine, British Medical Journal,* the *London Evening Standard, Scientific American, Boz* magazine and the *Biologist* for permission to include extracts from my articles for them. My thanks to the publishers of my books – including *Microbiology and Food* (1970), *Microbe Power* (1976), *Single Lens* (1985) and *BSE the Facts* (1996) – for some of the facts, figures and author's illustrations. Some of the key concepts in this book have been aired at meetings at the University of Cambridge and at Inter-Micro in Chicago.

I

Introduction

WELCOME TO THE NEW AGE OF BIOLOGY. The new millennium brings with it a fascinating array of new scientific techniques to control life, and the chance to find answers to age-old afflictions that burden us with illness. It will also need us to look at life in a new way– new philosophies that can help us make sense of what we are discovering. The molecular biologists have been retreating further and further into the structure of the cell, but what I want to do here is fit new knowledge into larger patterns. As the millennium draws breath we are drowning in data. What we now need are ideas to link our in-depth technical knowledge of life into a broader understanding.

Many people think that genetic engineering is to be feared. They said that about electricity, aluminium and motor cars, without which modern civilization would be crippled. We are told it can do things that no person could ever do, but so can a pair of scissors. It has been claimed that it gives us the power of destiny, but so do life-support machines. People argue that it interferes with nature, but we already do that with adoption and artificial insemination. We fear that it can exceed our human capacity to comprehend, but so do computers. People are anxious in case it creates unexpected problems and strange-looking mutants – but have you ever watched a Pekinese dog struggling for breath? What we need is sensible control. It was once the law for someone to walk in front of motor cars with a warning flag. Now we accept that people die and are wounded by cars every day, and that cars are polluting the atmosphere, wrecking villages and draining vital raw materials. Neither extreme can be justified. The lessons we must learn are that we must initiate proper controls over applied science – but these should not handicap the clear benefits at hand. Suffering people all over the world are on the brink of salvation through the good side of genetic engineering.

In this book I want to celebrate the supremacy of the cell. Looking at our cells will teach us much about ourselves. Many people have seen parallels between the behaviour of people and that of other animals. The science of sociobiology has offered fascinating insights into human nature, and now we need to understand where these behavioural patterns have their roots. In my view, the way we behave mirrors the properties of living cells. Once we understand how single cells interact, we can get to grips with the realities of human nature. At the root of it all lie the genes, the stored information within living cells that provides the key to make new life. Now that science is probing deeper into the molecules of life, we spend too little time fitting our new knowledge together. Facts are sterile until we assemble them into understanding. I'd like the new millennium to look at science as a holistic exercise, and begin to embrace great new ideas. The 2000s should open with the century of wisdom.

The 1600s saw the birth of modern science, and the acceptance of the microscope, the telescope and the principle of experimentation. In the 1700s, we saw the fashion for classification – of stars and planets, minerals and living organisms. Great voyages of exploration opened vast new empires of discovery and exploitation. During the 1800s, we were recognizing the structure of life, the nature of physics and the flowering of industrial chemistry. In the 1900s, we witnessed the exploration of the structure of matter: electronics, nuclear power, computers and a whole range of brilliant analytical techniques were unleashed. It brought awe-inspiring new insights into astronomy and how we could control chemistry and adapt it for our purposes. We began to analyse the structure of living matter, and to unravel the chemistry of life; for the first time we visited other worlds.

The new millennium heralds an explosive growth of scientific knowledge, but we often lose sight of what science really is. Much of what passes for science these days is technology. Now that we have new techniques for delving into the chemistry of life, and are discovering revolutionary facts about our genes, I believe that the overwhelming need is to unite it all into patterns of science that everyone can understand. Until now, science has been essentially reductionist. We have been probing into smaller and smaller details, often losing sight of the big picture. In the future, we need a holistic approach to unite disparate findings into the tapestry of truth. By gathering together the strands of discovery from many different fields we can begin to see how life works the way it does. This offers a new opportunity for wonder and sheer inspiration, for there is nothing more marvellous than the majesty of life.

Now we face a major problem. The public pay for science in two senses – not only does the public provide the money for science to continue on its course, but they are obliged to endure its vicissitudes. The public also pay for art and sport, for politics and economics, but science is uniquely resistant to perturbation by what the public want. The major enterprises of human society have to succeed or they will lose support, and if they are not mindful of what the public want there are forces available to bring about change. Not in the case of science. People simply endure it, sometimes revelling in its new insights, occasionally recoiling from the new threats we suddenly face. This distancing of the public from science is the result of an inability to follow what is going on. You know about the craters on the moon and the functioning of a digital display. You also know about living cells – but you couldn't draw one. When sport or microchips feature in a television commercial, they are properly portrayed. If it is germs or the cells of your skin that feature in an advertisement, then all we see are jabbering cartoons and gross caricatures of nature. Everybody is affected by germs, but few people have the least idea what they are really like. We are made up of living cells, yet few people seem to have any idea what one looks like.

The reporting of science does not always help. Science reportage does not simply contain mistakes, but layers of errors, each compounding the rest. The national newspapers feature reports about 'E-Coli' even though *E. coli* is the only right way to print it. The printing of the name as 'E-Coli' is no more correct than printing 'bill-clinton' or 'tony-blair'. Magazines write about *E. coli* O157, when that zero is actually meant to be a letter O. If the news on television carries an item about *E. coli* O157, you may hear them describe it as 'a deadly virus'. Even that is wrong. It is not a virus, but a bacterium. The reporting of a single disease germ brings layers of misunderstanding, each of which compounds the others. When there are so many errors in the media that inform the public, there is no hope of people understanding the issues. Until these matters are understood, there is little chance of the public beginning to influence science and science policy. All we have at present is a series of over-reactions to snippets of information. We throw up our hands in horror at the idea of a cloned sheep, but we have been creating new species for years. New germs such as *Cryptosporidium* and *Chlamydia* suddenly make news, but hardly anybody knows what they are.

Most popular science tries to skate around Latin names, but I am not abandoning them in this book. The garden centres of supermarkets are thronged with people mulling over *Metasequoia glyptostroboides* or buying plants of *Mesembryanthemum criniflorum*, just as children are thrilled by dinosaurs with names like *Tyrannosaurus rex* and *Therizinosaurus cheloniformis*, and yet there are no words so complex in my discussions of cells and how they act. In any event, today's rash of new diseases make many columns on popular medicine read like a gardening catalogue. Latin names are part of common culture.

This book looks at the achievements of the life sciences, and sets out some of the subjects that everyone will need to know in this new millennium. Everyone accepts that the public find it hard to keep up with science, but we should now face the fact that most scientists fail to understand the public. They act almost as though there is a right to carry on regardless, leaving the public to find their own way to catch up. Slogans take the place of facts. At a recent meeting, a science presenter reminded his audience that bringing science to the young is the most difficult task of all. That's absurd! Young members of the public are excited by science. A new comprehension underpins the way they look at life. They are familiar with the products of a scientific age (antibiotics and computers) and use science in their daily lives (from cash dispensers to mobile phones). They even joke about it. In the irreverent British comic *Viz*, the most iconoclastic of all youth publications, the character Johnny Fartpants has a gas chromatograph fitted by his father to warn of noxious gases before they escape and embarrass the family. In a *Viz* cartoon, a white-coated scientist is seen looking through his microscope on a table in the middle of a meadow, when a young on-looker enquires: 'Tell me, Professor, how long have you been working in this field?' Alternative comedians in the comedy clubs include jokes about genes and black holes in their patter, whereas science-based thrillers on television are increasingly popular with a young audience. Some of the greatest hit movies (such as the Jurassic Park films and the James Bond series) base their story lines on science. Since Thomas Dolby had television pundit Magnus Pyke intoning the words 'She blinded me with science' in a successful single, it has even been a feature of the popular music scene.

There has been a major collision between science and religion, notably in the United States of America, where the two are seen as irreconcilable extremes. This cannot be right. If religious fanatics insist that creationists are flying in the face of faith, then why should we not accept that – if God created everything – he created evolution? Many scientific sceptics insist that a knowledge of the intricacies of the living cell destroys the mystery and sense of wonder that a layperson might experience. This is not the case. Only when you get to grips with the mechanisms inside a living cell can you start to appreciate the marvels of life. The fact that the dazzling array of spring flowers and fluttering butterflies develop from such simple codes, and such basic chemistry, adds to the sense of wonder that all biologists feel.

I also seek to refute a growing belief in biology – that the cell theory is finished. 'The cell concept is now an empty shell,' said a recent publication, confusing its metaphors. 'The cell, as the centre of structure and function, is dead.' I believe the opposite: that we can only understand ourselves as manifestations of the cells of which we are composed. Everything we do is a reflection of the separate cells whose choreographed interactions make us what we are. Today, the cell is viewed as a dynamic chemical system. Youngsters learn of DNA and enzymes, but they see little of the intricacies of life. So much of modern science is nothing of the sort: it is laboratory technology, a very different discipline. The technicians whom we mistake for scientists are analysing the components within cells, and recording what they see, but this is no way to understand the wholeness of life in the round. That must be the task of the third millennium. Life is predicated on the activities and the internal decision-making processes of the cells that created our biosphere. They regulate it still. The single cell is a temple of pluripotentiality. Cells can build wings with which to fly, scavenge for sticks to make homes, construct stone walls for protection, manufacture delicate glass globes of infinite intricacy, and yet do not hesitate to sacrifice themselves wholesale for the good of the race. Unlike humans, they can regulate their rate of reproduction to the available food supply, and many can create a hermetically sealed, personalized 'space capsule' for long-term survival if their environment is disrupted. Of course cells are sentient. Some microbes can detect magnets or see where they are going.

We are seduced by the view that an amoeba is a simple creature, little more than a blob of plasm. In truth, its intricate life systems, undertaken by something so compact, connote hidden complexity. An amoeba has a head and a tail – confine it in a cul-de-sac and it turns round to find its way out. With two sticks of firewood and an elastic band, anyone can show how a human walks. It is harder to model the ability of a water-soluble amoeboid cell to feed and reproduce, find (and approve) a sexual partner, sometimes even to unite in vast numbers to form a macroscopic body that heads off looking for a place to breed.

This is how we may marry those two opposing concepts of the freshly fertilized egg cell, the zygote, as an ultimate microdot, a single seed for new life, and yet as the cell from which all later specialized cells can find expression. Multicellular organisms exist by potentiating specific functions in identifiable cell populations, while repressing those inappropriate to that site. Differentiation is the consequence of the development, in a pluripotential cell, of only those features that are functionally appropriate. This is the opposite of reductionism. Rather than retreating to the cell, I believe we should start with it and move outwards to multicellular magnificence. The capacity of cells to build communities, as people do, offers a truly holistic view of higher forms of life. By comprehending the cell, we can extrapolate from the model to embrace every multicellular species. In single cells we may model society. Every action of a many-celled organism is an expression of the functions of those single cells. Lust and pair bonding are mirrored when we see a grazing ciliate break off feeding to seek out, identify and conjugate with a chosen partner.

You are already blessed with immortality. The human species, *Homo sapiens*, does not die. The single cells live on as the germline: the ovum and the spermatozoon. Their survival in that bond which creates tomorrow's people is our own substance regenerated and living on when we are gone. There is nothing about the appearance of these cells to suggest that the two might fuse: that the globular cell might become infected by a swimming germ to initiate that process known as embryological development. We are the sporing phase on which the survival of these simple cells depends. If you fix on the germ cells, then that is *Homo sapiens*, floating about in brine as microbial species did a thousand million and more years ago. We humans are moribund, no more than senescent fruiting bodies facing an inevitable and distressing demise. People are the expendable husks which are supervened by the immortal germ cells. Such is the true nature of our kind. The cells matter more than us. If all human life ended tomorrow, our planet would hardly notice the difference and nor would a distant

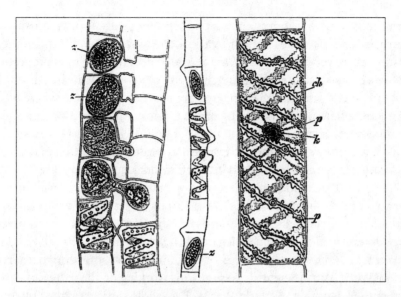

Sexual reproduction among single-celled plants. The aquatic alga *Spirogyra* (right) has attractive, spiral green chloroplasts which capture the energy of the sun. As this Victorian engraving clearly shows, the cells grow in chains, and during autumn two adjacent chains may conjugate (left). A tube forms to unite cells from each filament, and the male cells migrate through the tube to fuse with the female cells (far left). This shows that even these diminutive plants rely on sex to rearrange their genes.

observer. If the world of microbes became extinct, on the other hand, we would be unlikely to survive a year.

Living organisms have evolved a capacity to distinguish self from non-self. From the earliest and most simple forms of life, this ability is crucial for survival. This single characteristic, like reproduction, excretion and respiration, is a fundamental property that distinguishes the living from the non-living. Microorganisms contain genes that permit them to identify distinctions between self and non-self, and we see this property at work when the time comes to select a mate, for this is when they display an ability to reject unsuitable partners. Cells have a finely tuned capacity to recognize non-self, and the cells of the immune system can extirpate non-self life forms with vigour.

The animus of non-self rejection exists within all communities of living organisms. Creatures have a remarkable ability to recognize their relatives. We will see that bees can find their kin in the turmoil of a mixed community. Leaf hoppers that are morphologically indistinguishable detect tiny differences in stridulation that mark one from another. Instinctual rejection of non-self is one of the most fundamental properties of life: a unifying concept that stretches across biology. It is an area in which systemic theory and international relationships can meet, and where biology could offer an understanding of the origins of human conflict. I want to show that we can understand ourselves only as manifestations of the cells of which we are composed. Everything we do is a reflection of the separate cells whose choreographed interactions make us what we are.

The path to our modern understanding of the genes has a longer lineage than you might imagine. It began thousands of years ago, when ancient writers recorded unexpected patterns of inheritance and started to breed the domesticated species on which modern farming relies. The cell was discovered in 1663, and gene maps of chromosomes were being prepared before World War I. Scientists were already working on the nature of DNA before the beginning of World War II. The whole story is marked by controversy, by personalities who worked in defiance of their instructions, and sometimes under circumstances where there was little but animosity shown by the authorities. It is a tale marked out by personal rivalry as much as by pure science and the quest for knowledge. Yet the successes have been considerable – there have been many Nobel Prizes for the scientists who worked in the field, and sometimes the award was divided between colleagues engaged in similar quests (on one occasion to a father and son research team).

This emphasis on structure means that today's schoolchildren learn little of the nature of life. Modern youngsters have lost out on knowledge of the plants and animals that surround them. The old subject of 'nature study' was transmuted into 'environmental studies' and, as the new molecular biology has filtered through to the school syllabus, a full understanding of life in the round has been abandoned. One of the simplest and most vivid experiences you can offer a child is the sight of living cells under a microscope. A droplet of pond water is a most astonishing spectacle. Yet, for all our trumpeting of our ability to penetrate inside the mechanisms of a living cell, we do little to reveal to people what cells themselves are like. The living cell is the fundamental concept that underpins life on earth, and the variety of cells is breathtaking. People need to have access to this key and important truth; it is time to liberate cells and present them to a modern audience.

Now I want to go one stage further. This book sets out a holistic view of life, and asks that we set our new knowledge into context. The truth of the matter is that all life shares common propensities. In single cells we see how they recognize each other and communicate; in humans we see the same phenomena reflected in tribalism and conflict. Our continuity with all nature lies encoded in our genes.

Genes are far from 'selfish'; they are the servants of the communities of cells. To imagine that life exists to serve genes is like pretending that holidays are invented to give passports a break. A ruling gene would make mere robots of us all, and the variety of life is more than such simplistic ideas can explain. In any event, this simplistic notion of the 'selfish gene' does not reflect observable reality. As I was writing this book, there were press reports of a brave cat who fought her way through flames in a house fire to save her kittens. She lost almost all her fur in the process and was so badly burned it seemed unlikely that she'd recover, though she did. What would a 'selfish gene' theory have to say about that? The cat would have done better by staying out of the fire and having more kittens later. Her litter would have perished, thus eliminating a race of non-fire-retardant cats from the population, but by trying to rescue them she runs a risk of perishing with them in the fire, thus eliminating all their gene pool. The bravery she showed – if a genetic imperative – would doom her to extinction, for timid cats who avoided all danger would be the only ones selected. Exponents of the selfish gene will argue that it was the genes that drove her to take dreadful risks, but that implies a magical additional dimension. By not taking risks, her genes would have survived intact; the genes she is helping to survive by running into the fire were those of her mate. To argue that these unrelated genes influenced her actions would be to assume some form of behavioural contagion for which there isn't the slightest evidence. If the genes were selfish, they'd have the parent run and hide; the offspring might die, but the parental genes would in this way be conserved for the future. By running the risk of rescuing the young from a fire, all may die – and the genes would be destroyed with them.

Populist scientific theories are often rich in resonances of the culture of the time. 'Selfish genes' became fashionable when a self-centred ethic prevailed in personal relationships. Their proponents show that computer models support a selfishness principle, but that should not lead you to believe that the theories are necessarily right. Computer models are nothing more than that – human constructs that act as theoretical models. They are not predicated on the real world. Sometimes they are of relevance, but the mere fact that a model fits the facts is not evidence that it provides the right explanation. We will see that altruism is a key motivational factor in life, and I believe that human nature is rich in altruistic impulses that the modern world tends to decry.

We will also review evidence in this book for a nineteenth-century proposal: that modern highly evolved cells have been produced as earlier cells began to coexist within the same envelope. The tiny mitochondria which act as the cell's power-house still contain traces of primeval genes – but most of them have been lost. Were these 'selfish' genes as some like to proclaim, they would have ensured that they propagated, not degenerated. The cilia with which some cells swim may once have been independent bacteria, which later colonized cells and devoted their abilities to the functioning of a more complex structure. Were they possessed of 'selfish genes', the genes would have survived at all costs. In fact they have faded away over millions of years of evolution. If there is any selfish component of life it is the living cell, at least one of which always passes on to the next generation. We exist to propagate these cells. The gene is a part of an organism's armoury, and not its master.

Life does not devolve on discrete and identifiable chemical reactions, but exists in the full-flowered majesty of a living cell. It is communities of cells that come to make a human, and it is the ways of these single cells that rule our nature. Humans have longed for immortality since time immemorial, and now we can see how we have always been immortal. Each person owes his or her identity to the living cells that fused to create a new generation. This is what becomes a new person, and in this way – for the reproductive cells – there is no death. I believe that our behaviour is nothing more than a reflection of the proclivities of each cell within our bodies. Our human nature is rich in the resonances of their behaviour; only when we come to grips with the way cells work can we start to understand our own humanity. The cell theory is far from dead. I believe it is only just fit to come fully to life.

2

Discovering the cell

THE STORY OF THE CELL began in April 1663, with a young experimenter named Robert Hooke (1635–1703). He was working with his new microscope at the command of the Royal Society of London. Hooke had been born at Freshwater on the Isle of Wight and was educated in London and Oxford, where he became assistant to Robert Boyle, the pioneering chemist. While still in his twenties he moved with Boyle to London, where the newly formed Royal Society appointed him as their demonstrator. On 25 March 1663, he was charged with bringing in a new microscopic demonstration every week. From the very first demonstration, Hooke gave presentations that set in train the modern era of science.

He was an industrious young man, responsible for some fine architecture and is widely known for 'Hooke's law', which explains the stretching of a spring. Two weeks after he was first charged he brought in a demonstration of the appearance of moss under the microscope. He published an engraving of the images in his influential book *Micrographia*, in 1665. There is a striking feature in the drawings he made of the leaflets of the moss: each showed with perfect clarity that it was composed of tiny cells, each delicately fitting to its neighbour like bricks snugly fitting in a wall. The following week, on 13 April, he returned with a further investigation of these little structures. This time he had prepared a fine section of cork from a bottle stopper, and showed these little brick-like structures in greater detail. To Hooke they seemed like tiny square rooms, and this was why he termed them 'cells'. This is the term used in modern biology.

The discovery of living cells. On the afternoon of 8 April 1663, Robert Hooke demonstrated a specimen of wall moss to Fellows of the Royal Society in London. It reveals the presence of living cells, fitted together like tiles. This epoch-making view of nature also gives us an immediate impression of the true size of cells, relative to a little moss with which most people are familiar.

In due course, his work was compiled into a book, *Micrographia*. It is a large volume, measuring thirteen inches in length and weighing three pounds. The book is printed with a publication date of 1665, but it was actually released in October of the previous year. It became an instant best seller, as Hooke had clearly planned. Before he started writing, he did some market research and made sure that nobody else was likely to bring out a rival book which might damage his own chances. In the book he had a number of highly vivid images (including studies of painfully familiar subjects including a flea and a louse on pages so large that they had to be folded twice to fit in the covers). One of the purchasers was Samuel Pepys who had owned a microscope since he joined the Navy Office. In his *Diary* he said that *Micrographia* was better than any other book he had purchased. He sat up half the night reading it. The book was so successful that a second edition was published in 1667. The engravings were brought out of storage 70 years later and reprinted under the title of *Micrographia Restaurata* with further editions in 1745 and 1780. Facsimile editions of the original book have been published in paperback in the twentieth century,

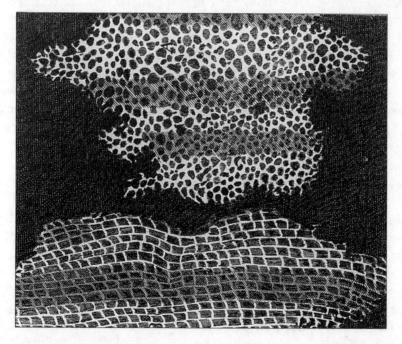

The naming of the cell. This is the study of cells that gave them their modern name. It shows a fine section of cork, cut by hand using a shaving razor. The name was coined in April 1663 by Robert Hooke, who was demonstrating his new microscope to the Royal Society of London. Hooke's fine engraving was published in his book *Micrographia*, dated 1665.

making *Micrographia* one of the most successful scientific books of all time.

Shortly after the book had been published, and when it was at the height of its popularity among the *cognoscenti* of the British intelligentsia, a traveller from the Netherlands sailed up the Thames to visit business contacts in London. He was Thonis Leeuwenhoek (1632–1723) whose father-in-law was a textile trader who had lived in Norwich. Leeuwenhoek used a lens to assess the quality of cloth, and in *Micrographia* appeared some beautifully executed plates of cloth under the microscope. They have a stunning, three-dimensional quality, very much like modern scanning electron micrographs. To the curious mind of Leeuwenhoek, the descriptions in Robert Hooke's book were a crucial stimulus. In the Preface, Hooke described exactly how to make a little hand-held microscope with a ground glass bead for a lens. The magnification, said Hooke, was far greater than that obtained with the more conventional compound microscope which he preferred to use. The problem was that they had to be held very close to the eye, and were therefore inconvenient.

Over the following years, Leeuwenhoek began to make experiments of his own. In time he perfected Hooke's design, polishing his lenses from melted beads of glass. To show his prowess, he re-examined some of the specimens described by Hooke, just to show that his observations were of higher quality than those of the English pioneer. He relayed his findings to the Royal Society in a letter dated 1 June 1674. His great 'eureka' moment came late in August that year. Leeuwenhoek took a boat trip across a lake, Berkelse Mere, where the water was clouded with mid-summer growths of microbes. Local legend held that this was honey dew, caused by condensation during the chilly evenings, so Leeuwenhoek collected some samples in a glass phial and took them home. The next day he set to work, examining the material under the best of his hand-made microscopes. What he saw was to change our view of nature forever. In front of his astonished gaze were a myriad tiny living organisms. They swam about, twisting and turning in front of his eyes, while others lay still, their inner workings glistening in the light. He watched this extraordinary sight until he was exhausted. As he could not draw – Leeuwenhoek often remarked on his inability to capture a likeness with a pencil – he employed a limner to observe the specimens and make detailed drawings of what he saw. So excited was the young artist at the sights that faced him, that he was regularly reminded to stop gazing and get on with the work. Leeuwenhoek and his (unnamed)

A colony of rotifers portrayed in 1905. Rotifers were vividly described by Leeuwenhoek. These delicately constructed organisms are composed of many cells, though not too many. Some are made of no more than 100 cells. This type is commonly found in ponds and ditches, and extends fine arms with which to catch its prey. These organisms are characterized by a conspicuous capsule of jelly, into which the rotifer can retreat if threatened by a predator.

assistant were the first people in the world to study the world of micro-organisms. Single-handedly, he discovered protists and algae, rotifers and polyps, and then went on to describe bacteria. Leeuwenhoek was the person who gave the world our new science of microbiology.

The cells he described were not the inert structures observed by Hooke. For Leeuwenhoek, the new world he discovered was teeming with micro-scopic organisms. As an adult he used the forename of Antony, and as he became better known he added a 'van' to confer a greater sense of dignity to himself. He had many visitors who wished to see more of his work, includ-ing the rulers of Britain and Russia, and was elected to membership of several of the leading scientific academies. Antony van Leeuwenhoek was the first microbiologist in history, and the way that we conduct our experiments today owes much to the insights that he pioneered. He was wonderfully single-minded: self-taught and independent, he declined

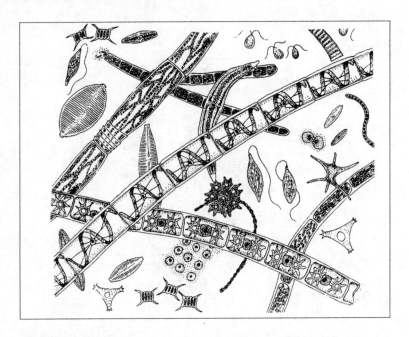

Living organisms from a sample of pond water. This is the kind of sample which Leeuwenhoek first observed in the 1670s. Several strands of green algae lie across the image, and some of the single-celled organisms (near the centre) have whip-like flagella with which they actively swim. Even the oval-shaped diatoms and desmids are in continual movement, gliding across the field of view. To Leeuwenhoek the sight was entrancing, and he continued to study specimens until he was exhausted.

to teach students and never made public presentations. His view was a simple one – people were unwilling to accept what he saw, and he had better things to do than waste his time explaining. 'I know I'm right,' he said, and he was.

During the later part of the eighteenth century it became fashionable to purchase small microscopes to examine the wonders of pond life. *Hydra* became a popular topic for these hobbyists. It is a delight to observe their tenuous tentacles delicately expanding through the water and capturing water fleas, using fantastically complex cells that eject barbed spears at high speed and anchor the prey on threads as strong as steel wires. The body of the animal is about a centimetre (half an inch) in length, and it is composed of a hollow tube made of just two layers of cells. *Hydra* was studied by Leeuwenhoek and drawn by his limner in December 1702. During the

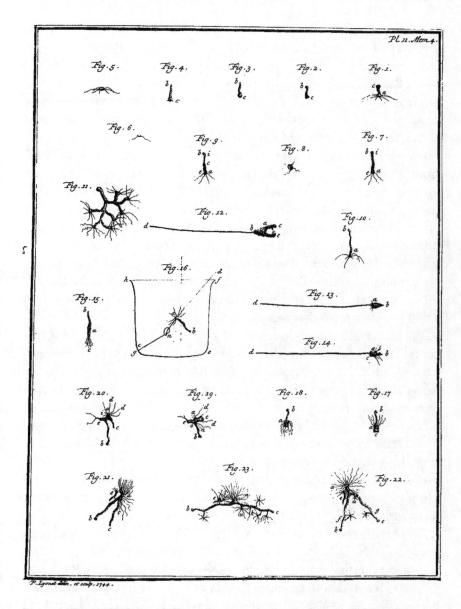

Trembley's experiments with grafting. The freshwater polyp *Hydra* is a miniature sea anemone. Abraham Trembley carried out numerous experiments on this organism in the eighteenth century, and showed how transplantation and tissue grafting were possible. His experiments also showed that *Hydra*, which has no eyes, could none the less respond to light. He illustrated his experiments with fine engravings.

1740s, a young philosopher named Abraham Trembley (1710–84) turned to *Hydra* as an object for study. At the age of 30 he had been appointed tutor to the two young children of Count Bentinck of The Hague in the Netherlands, and he used the microscope in his classes. Trembley was fascinated by these small creatures. He observed them and drew them accurately, and then set out on a series of dramatic experiments. Even today, they seem extraordinary for their time. Trembley carried out pioneering tissue grafts. He showed how *Hydra* could regenerate itself from small fragments of tissue. He showed how tissues could be stained, how an animal without eyes could respond to light, and even described the cytoplasm of which each cell is made. Since then, Trembley has been largely forgotten. His name is absent from most biographical reference books, for instance. His intriguing experiments were published in 1744 under the title *Mémoires d'un genre des polypes d'eau douce* in Geneva, and were recently published in an English translation. It is a beautiful book, filled with remarkable experiments. In this work you can find many hints at the great future era that awaited cell biology.

The increasing interest in *Hydra* led to the development of the first successful bench microscopes. Microscopes, as a gadget or a novelty, had been available since the 1600s. In skilled hands they had been put to good use by Robert Hooke and several other pioneers. The high-power microscopes were made of brass, but were still little more than curiosities. The low-power microscopes had bodies made from wood and cardboard, focused by a crude clamp on a pillar – not the kind of design you could upgrade to something useful for serious research. *Hydra* changed all that. Trembley used to observe his specimens with a hand lens. Often he placed them into a little pool of water trapped in the palm of his cupped hand, manipulating them with a bristle.

To observe this delicate little animal under a bench microscope, you needed a metal instrument that could be focused gently, without the jarring vibrations that would make the *Hydra* contract into a ball. It necessitated the design of a stage that was wide enough to support a watch glass holding the specimens. The instrument also had to offer a reasonable level of magnification, so that details could be observed.

One of the amateurs who studied *Hydra* was Henry Baker (1698–1774) of London. Baker was Daniel Defoe's son-in-law and he was preoccupied with the issues of biological philosophy which still concern us now. He outlined his ideas in some doggerel poetry published in 1727. It was described as *A Poem to Restrain the Pride of Man,* and in part this dreadful poem ran as follows:

> Each seed includes a plant: that Plant, again,
> Has other Seeds, which other Plants contain:
> Those other Plants have Seeds; and Those,
> More Plants, again, successively inclose.
> Thus, ev'ry single Berry that we find,
> Has, really, in itself whole Forests of its Kind.

This poem was satirized in 1733 by Jonathan Swift, who was a prolific critic of the science of his day. Swift's satire was deliciously pointed:

> So naturalists observe, a flea
> Hath smaller fleas that on him prey;
> And these have smaller fleas to bite 'em
> and so proceed ad infinitum.
> Thus, every poet, in his kind,
> Is bit by him that comes behind.

In more recent years a simplified version of that poem has become widely quoted:

> Great fleas have little fleas upon their backs to bite 'em,
> The little fleas have lesser fleas, and so ad infinitum.

Plenty of people know this last version, but few of those who quote that droll little couplet realize that it arose from the writings of Jonathan Swift – or that it reflects the weighty preoccupations of his time. Henry Baker ended his poem with what he called an 'amazing thought', concluding that Adam's loins must have contained 'his large posterity, All people that have been, and all that e'er shall be'. This was the continuing conundrum of biology – how could the entire essence of a creature re-emerge in the next generation? Some of the earliest microscopists had imagined that they could observe a microscopic person inside a sperm cell, and this – known as the *homunculus* – was the only way that anyone could envisage inheritance. Baker's poetry was a florid attempt to address the mystery and, meanwhile, it bequeathed to us one of the little jokes that scientists like to pass on down the generations.

Baker's interest in observing *Hydra* eventually took him to John Cuff, a London instrument maker. Could he produce a brass instrument, which was (in Baker's words) free from the 'jerks, which caused a difficulty in fixing it exactly in focus'? Cuff announced his new design in 1744, and this new microscope had many of the features of a modern research microscope – a milled wheel for focusing, a solid stage mounted on a solid body and a means of directing light up through the stage. Other manufacturers followed this lead, including George Adams who made surely the most grotesque microscope in history. It was designed for King George III in the early 1750s and was a massively ornate instrument made from solid silver. The British royal family showed a continuing interest in microscopy, which remained a hobby for the educated classes for over two centuries. Edward Elgar composed his greatest music with a magnificent brass microscope standing nearby, and his collection of slides is preserved to this day in his little cottage overlooking the Malvern Hills in the English Midlands.

In the following years many pioneers peered curiously through primitive microscopes in search of an answer to the conundrum posed by Baker. One was John Ellis (1710–76), an Irish-born British government official based in Florida and Dominica, who was an enthusiastic microscopist. He claimed to have seen tiny progeny lurking within the microbial cells. If he looked closely, why he could see microscopic future generations lurking within those little bodies. To him, life was like the layers of an onion, with each generation (just as Baker had proposed) enveloping the generations yet to come. The idea was disproved by a largely forgotten Swiss geologist, Nicolas de Saussure (1767–1845) of Geneva. His painstaking observations revealed the truth: microbes could reproduce by dividing in two. Ellis would have none of it – if microbes ever divided in two, he insisted, it was because they collided accidentally and broke apart. Saussure's findings were mentioned to an Italian eccentric, Lazzaro Spallanzani (1729–99), who was implacably opposed to Ellis and his school of supporters.

Spallanzani made careful observations of his own. By using a fine hair, he managed to separate a single cell into a droplet of water under his microscope, and settled back to observe what happened. He saw it divide in half, grow again, and divide once more; and he repeated the experiment time after time. Spallanzani was a theatrical and flamboyant character, who became a priest in order to obtain job security but secretly set about questioning every shade of current belief. Perhaps his greatest contest was with an English priest, John Needham (1713–81). Needham lived in France for most of his adult life, and propounded a formal theory of spontaneous generation in 1749. He even claimed to prove it by experiment. Needham kept mutton soup in little bottles and watched how microbes appeared, as if by magic, as the days went by. He sent his descriptions to the Royal Society, who were greatly impressed at his clarity of vision. Leeuwenhoek had concluded that microbes were descended from other microbes many decades earlier, yet Needham seemed determined to stand the idea on its head. He was quite convinced that microbes formed by condensation out of soup, and that was why stale soup was full of microbes. In his experiments, he poured samples of fresh broth into bottles and corked them shut. Within days, when he examined drops under a microscope, he saw that they were swarming with bacteria. Just to make sure he was right, he heated some of the bottles of soup in glowing cinders to kill any surviving microbes – and found that they still produced a crop of microbes after a few days. He repeated the experiment with soups made from plant matter, with the same result. This was enough to convince him that spontaneous generation was a fact.

Spallanzani was incensed at this assault on reason, so he set out to discredit the English cleric. He reasoned that the soup samples had been contaminated by bacteria: perhaps they had gained contact through Needham's handling as he corked them, or possibly the heat had been insufficient to kill bacteria already present in the soup. The experiment that he designed was thorough. Spallanzani set up a range of glass bottles filled with mutton soup, and also with other nutritious infusions like those used by Needham. Rather than seal them with a cork, he melted the neck of each bottle so that microbes from the environment could not possibly contaminate the contents. Then he set the bottles to heat in a bath of boiling water. Some he left boiling for a few minutes, and others he left in boiling water for hours on end. At the same time – by way of controls – he put aside an identical range of bottles which he corked just as Needham had done. When he uncorked the bottles, he found they were rich with a microbial growth. Little wonder that Needham had found his soup samples to be teeming with bacteria. Then he broke open the bottles whose necks he had fused in the flame. Those that had been heated for a long time were sterile, and completely free of any form of life. One or two of the sealed bottles that had been in the boiling water bath for a few minutes were contaminated, and Spallanzani realized that some microbes could withstand considerable heat. This, he felt, proved two points at once: Needham's flasks grew microbes because his sterilization was inefficient. He also showed that even boiling might not be enough to kill the most resistant microbes.

There was no mistaking the results. At least, that is how it might seem – but Needham had obtained a patron in the form of a florid French aristocrat, George-Louis Leclerk, Comte de Buffon (1707–88). With Buffon's support, he considered the results and retorted that the heat Spallanzani employed had destroyed the 'Vegetative Force' necessary for spontaneous generation to take place. Needham had been right all along: it was just that we now knew how easily the Vegetative Force can be destroyed by undue heat. This new idea of Vegetative Force was soon widely discussed across Europe, and seemed to have put paid to Spallanzani's theories. Needham was fêted across Europe, honoured by the Academy of Sciences in Paris, admitted as a Fellow of the Royal Society in London, and he and Leclerk continued to proclaim their views to everyone who would listen.

Spallanzani set out to disprove their claims, and he did so in a carefully considered series of experiments. Once again he took a huge range of bottles, and treated them exactly as Needham had done. Some contained mutton soup, others contained nourishing vegetable infusions; each one was set into a scrupulously clean bottle fitted with a fresh, new stopper. Some he set into the water-bath for a few minutes, others he left for an hour, and several others he left simmering all day. They were then left to cool, and were examined a week later. Spallanzani argued that, if the Vegetative Force existed, the longer his broths were heated, the more the Vegetative Force would have been destroyed. If it existed, the flasks boiled for a few minutes would have the greatest extent of microbial growth whereas those left on the heat for hours would have none. The results were quite clear. Spallanzani found that the extent of contamination was completely unrelated to the length of time that the flasks had been left boiling. Clearly, the growths were caused by microbes getting into the brews, and not by any inexplicable Vegetative Force. In the hope of clinching matters, Spallanzani tried perhaps the bravest experiment of all. He took a range of dried beans and seeds (the kinds used to make the broths) and roasted them on an iron griddle. This would surely destroy the last traces of a heat-sensitive Vegetative Force. He added to this some freshly distilled water, and left the broth made from the scorched remains to brew. After a few days, the resulting infusion was teeming with microbes. In this single demonstration he felt confident that he had proved that it was contamination that lay behind the phenomenon, and that the existence of Needham's Vegetative Force was surely disproved.

Needham fired a final salvo. The Vegetative Force needed air at normal pressures if it was to function, he proclaimed. When Spallanzani sealed up his glass flasks, he damaged the air. Needham could prove this, he said, because if you sealed up a bottle as Spallanzani had done you could hear the hiss of air into the bottle as you cracked open its neck. It was low air pressure that was the problem, he insisted, and he was right after all. Spallanzani carried out the test that Needham described, and found that indeed there was a hiss of air when he opened one of his sealed flasks. A candle flame showed that the air was rushing into the flask as it was opened, so clearly it was at lower pressure inside the closed container. Could it be that Needham had been correct? The point had to be addressed, so Spallanzani took a series of glass flasks and filled them with broth, just as before. This time, however, he drew the necks out in the flame until they were almost closed – but not quite. Then, fleeting contact with the flame melted the neck and hermetically sealed each flask. After treating them as before, he cautiously opened the flasks and observed what happened to the candle flame. This time the candle flame was blown away from the flask – clearly, the air trapped inside was at slightly higher pressure. With this final demonstration, Needham's obsessions were finally laid to rest. Spallanzani had disproved spontaneous generation.

As the nineteenth century dawned, an interest in life in all its forms was spreading widely through informed society. Great scientific expeditions were under way, and one of the adventurous young scientists of the day gave us the next important coinage in the study of microscopic life – the cell nucleus. The discovery that cells contained a nucleus was made by Robert Brown (1773–1858). He was a Scottish doctor who had travelled to Australia in search of new species. Brown worked for Sir Joseph Banks, who had travelled round the world with Captain James Cook on the Endeavour in 1768–71. A further voyage of discovery was planned with the 27-year-old Matthew Flinders in charge, and Brown became part of the team. In his first three weeks in Australia, Brown described 500 plants – almost all of them completely unknown to science. As he studied them with his microscope, he laid down some important new ideas. For instance, he realized that you could use the structure of pollen grains to help work out the classification of flowering species, a technique widely used to this day. It was also Robert Brown who observed the ceaseless jostling movements of tiny particles, a movement that reflects the agitation of molecules within a fluid. In due course this became known as 'Brownian movement' though it was not fully understood until Albert Einstein took the matter in hand and solved the problem mathematically in 1905.

To this day, there are people who doubt what Brown could really have seen with his little microscopes. Just as in the Leeuwenhoek instruments, there was a single tiny lens for magnifying. To many people it seems impossible that you could see much detail with such a primitive optical system. This scepticism does not stand up to scrutiny. I have recreated several of his observations using the original instruments and the view that they provide is surprisingly clear. Brown's microscopes were made by two manufacturers: the firms of Bancks (a father and son concern) and Dollond (a family of instrument makers). They are carefully crafted microscopes which fitted

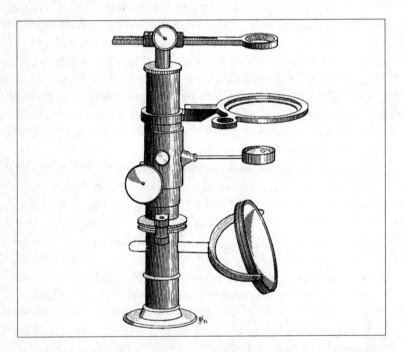

The microscope that showed us the nucleus. Robert Brown, a Scottish doctor and explorer, used this microscope to study the cell nucleus. He coined the term in 1828, after observing nuclei in orchid cells. Brown also studied Brownian movement, the ceaseless agitation of tiny particles related to the movement of molecules. People have doubted whether he could see such things, but I repeated his experiments and showed that he certainly saw what he claimed.

into a mahogany box when not in use, and descended from the design by John Cuff. With a mechanically operated stage, controls for coarse and fine focusing, and even a condenser mounted below the stage, these microscopes are still a delight to use. Their construction is echoed in the way modern optical microscopes are made. Robert Brown stumbled upon the importance of the nucleus while working on orchids. He teased out thin layers of orchid tissue for examination under his microscope, and wrote notes on what he saw. They were privately printed in 1832, and in a note Brown describes how this crucial discovery dawned on him:

In each cell a single circular areola, generally more opaque than the membrane of the cell, is observable … only one areola belongs to each cell. This areola, or nucleus of the cell as perhaps it might be termed, is not confined to the epidermis …

He went on to use the term exactly as we would employ it today: 'The nucleus of the cell is not confined to the Orchidae, but is equally manifest in many other Monocotyledonous families.' Within a few years, it was realized that all normal cells have a nucleus, and this important coinage has underpinned modern biology. Although this was the moment when the term was introduced into science, it was not the first time that nuclei had been recorded. Leeuwenhoek had featured them in his own publications, though he had not noted their widespread occurrence in nature – nor did he give them a name.

Leeuwenhoek and Brown were dour and self-contained individuals, and you might imagine that this was a characteristic of all pioneering biologists. Not so. Although a devotion to the task in hand might seem to be just what a scientist needs, there were highly flamboyant characters who helped to uncover the nature of the living cell. One such was Matthias Schleiden (1804–81) of Jena, who first put forward the modern-sounding idea of form-building forces which, he felt, explained the growth of inanimate crystals in the same terms as the formation of a bodily organ in a developing embryo. Such 'force field' theories have become popular in the closing years of the twentieth century, and few people realize that they owe their origin to an eccentric German botanist in the mid-nineteenth century. Schleiden was born in Hamburg in 1804, the son of a leading physician. He began his studies as a lawyer and took up the practice of a barrister in his home town. It was not a success. He was moody and so unreliable that he could not build up his business. In a fit of utter despondency he took a gun, pressed it to his forehead and shot himself. The bullet did not penetrate, and he made a full recovery in remarkably good time. Equipped with his degree in jurisprudence he decided to become a physician, so he qualified in medicine and then studied natural philosophy which brought him his third doctorate. In 1850 he was appointed Professor of Botany in Jena, and it was there that he began to review Robert Brown's discoveries. Schleiden recognized something that had not dawned on Brown: not only was the nucleus a widespread feature of living cells, it was also of crucial importance. Some of his conclusions were wrong (for instance, he believed that the nucleus dissolves away in mature cells) but in other respects he made great conceptual advances. He recognized that a mature plant is a community of cells, and this opened the door to our modern view of life.

His theories were extended by Theodor Schwann (1810–82) who was born in a village in Rhenish Prussia where his father was a bookseller. Schwann studied medicine, and moved to take up a professorship in Belgium in 1839. He recognized that decay and fermentation were caused by microorganisms, and he also discovered pepsin in the gastric juices. He then took Schleiden's work and extended it to become a general theory of life – that all organisms are composed of one or more cells. He also described cell division, which was itself an important advance. From this time onwards, the cell theory has always been thought of as the theory of Schleiden and Schwann.

Their views were wrong in some fundamental ways, however. Both of them imagined that cells could arise from the condensation of moisture, and Schwann shared the view that nuclei are transitory – they appear when the cell is young and vanish as it ages. It is true that you may find nuclei missing from dead or dying cells, but only one type of familiar cell lacks a nucleus altogether. This is the mammalian red blood cell or erythrocyte, the most common cell in the human body. Many creatures (like frogs and toads) have oval red cells, with a prominent nucleus at the centre, but the typical mammalian erythrocyte loses its nucleus as it matures, exactly as Schleiden and Schwann proclaimed. Our own red cells, for instance, are flattened discs with hollow sides, as though marking the missing bulk of a nucleus in the middle.

In 1869 a Swiss chemist named Johann Friedrich Meischer identified a substance called DNA within the nucleus, though its significance was yet to be discovered. Science moved nearer to understanding the importance of the cell nucleus through the work of a Swiss zoologist, Rudolf Kölliker. He

Living cells in the web of a young frogs foot. Until the twentieth century it was common to find a frog plate with a microscope – a device for temporarily holding a living frog so that the astonishing complexity of its blood circulation could be observed. This Victorian engraving shows fine capillaries containing blood cells (unlike those of humankind, frog red cells contain a nucleus). There are also three dark-coloured cells in this field of view. These, the melanocytes, expand and contract in response to light levels, and help the frog remain camouflaged.

did not altogether rule out the formation of cells by some form of 'condensation', but he regarded it as a matter of little consequence. Instead Kölliker expounded the importance of the nucleus to the cell, and the importance of cell division. His views were further advanced by Ludwig Virchow of Berlin who studied medicine and advanced the cell theory as the fundamental property of all life. Virchow was a pioneering pathologist, and introduced an idea that disease was caused by cells. Curiously, however, when Louis Pasteur popularized the germ theory of disease, Virchow was disinclined to accept it.

The enduring mystery of inheritance was to be resolved by two groups of biologists. The first were those who were observing the cell, and it was through observation that the workings of life began to emerge. Eduard Strasburger, a Professor at the University of Bonn, published a book on cells in 1875. In it he set down the old idea that the nucleus of a plant ovum dissolved after fertilization, and new nuclei form as a result of condensation. However, just five years later there was a new edition of his book – and now he had changed his mind. In this revised account, he laid down a fundamental principle of cell biology: all nuclei arise as a result of division of a

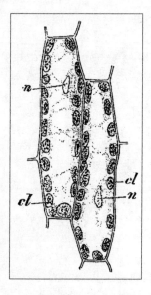

How Strasburger viewed the cells of moss.
Eduard Strasburger, a noted German botanist,
studied moss cells in the nineteenth century. This
figure appeared in his great *Textbook of
Botany*. Each of the cells contains many green
chloroplasts filled with chlorophyll (labelled 'c')
and a single nucleus ('n'). These two cells can be
compared with Hooke's studies (see page 13) to
give a clear idea of the real size of these structures.

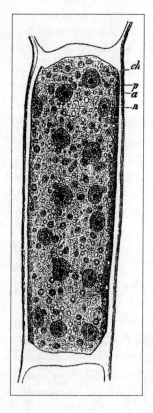

Cells can contain many nuclei. The aquatic alga
Cladophora is popularly known as witch's hair or
blanket weed. It grows as chains of cells forming
long hair-like green strands in water. As
Strasburger correctly deduced, each cell contains
a number of separate nuclei ('n'). These cells also
contain pyrenoids ('p') where sugars and starch
are formed during photosynthesis. Each cell is
about as broad as a typical human hair.

pre-existing nucleus. They do not dissolve away, and they certainly do not form from condensation or any other such process.

By this time, several scientists had made careful observations of dividing cells, and Walther Flemming (who worked at Prague and later at Keil) published the first clear account of a dividing nucleus. He even noted ribbon-like structures which appeared within the nuclei during cell division, and showed how they were produced in matching pairs. In 1888 they were given the name by which we know them today: chromosomes. These are the structures inside the nucleus where the genes are hidden. We will return to the research in these crucial and fascinating fields in Chapter 8.

The second group were the experimental biologists who observed how organisms behaved. In 1886, a Dutch botanist, Hugo de Vries (born in Haarlem in 1848), was given some plants of evening primrose which he planted in his garden. Much to his surprise, the seeds that they produced grew into a new generation of plants with markedly different appearances. Some were dwarfed, others had different-shaped leaves. He began to investigate the phenomenon, and concluded that species could change through sudden and unexpected mutations. Unknown to de Vries, a Moravian monk, Gregor Mendel, who was 26 years his senior, had experimented by cross-breeding peas. Mendel stumbled on the fact that hereditable characteristics could be understood as single, separate entities – he had stumbled on the existence of genes. He published his findings in an obscure journal and they were generally ignored. In more recent times, we have examined his results more closely, and it is now clear that he cheated in publishing his results. They are far too good to be true. Natural systems exhibit subtleties of variation which are absent from Mendel's work. It is now clear that he knew he was right, and that inheritance was based on these indivisible units we now call genes, so he didn't take the trouble to do the experiments he should have done. Instead, he made them up and published them anyway.

A Danish-born agriculturist, Wilhelm Johanssen, showed that the nature of an organism is defined by its genes. He noted that, if bean plants were grown in poor conditions, both they and their seeds were smaller than usual. Plants grown in well-watered and highly nourished conditions became large and verdant, producing beans that were far larger. However, even after many generations, the inherent characteristics of the species did not change. Small beans and large beans planted together gave rise to plants that were similar in size and productivity. Johanssen drew a distinction between the genotype (the inherited characteristics of a species) and the phenotype (the way in which those characters are expressed). And then came the breakthrough – the discovery of chromosomes as the bodies that contain the genetic code. They were first observed by Karl Nägeli in 1842 (see page 158) and beautifully drawn by Karl Rabl in 1887. They gained their present-day name from Waldeyer in the following year. Once chromosomes had been observed it began to seem clear to many biologists that – as they were passed from one generation of cells to the next – they must have a close connection with the transmission of inherited characteristics. The most fundamental of all inherited characteristics within a species is sex, of course, and in 1906 the determination of sex by sex chromosomes was discovered. An influential American investigator, Edmund Beecher Wilson (1856–1939), recognized that the chromosomes were responsible for conferring sex upon a new organism. Nettie Stevens (1861–1912) at the Bryn Mawr College in Pennsylvania described the X and Y chromosomes in 1906. She showed that the X chromosome confers femininity, whereas the Y chromosome is the male sex chromosome. From this moment, the search began for other hereditable characteristics. Scientists probed deeper into the nucleus, trying to relate inherited characteristics to the structure of the chromosomes.

The groundwork for research into the nature of chromosomes was laid by a tireless and diligent American, Thomas Hunt Morgan (1866–1945),

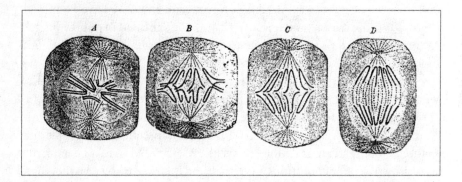

How the pioneers recorded the division of cells. Carl Rabl made this drawing of dividing cells in 1887. He showed how the ribbon-like chromosomes formed a group near the middle of the cell ('a') and then were pulled towards the two ends of the cell by fine threads which form during division. Rabl's drawings show exactly how the chromosomes split in two and are drawn apart as the threads of the spindle contract. This is how DNA is distributed between the daughter cells.

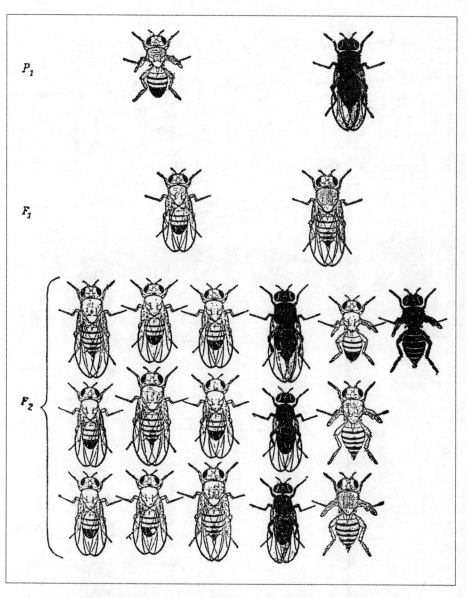

Classic cross-breeding experiments with the fruit fly. The first species to have its genome investigated was the fruit fly *Drosophila*. Cross-breeding can reveal mutants, and a study of their abundance can be traced to the relationships between the genes involved. This original study by Thomas Hunt Morgan shows how a pair of adults can give rise to offspring of two similar types. The progeny of this generation includes black and wingless mutants.

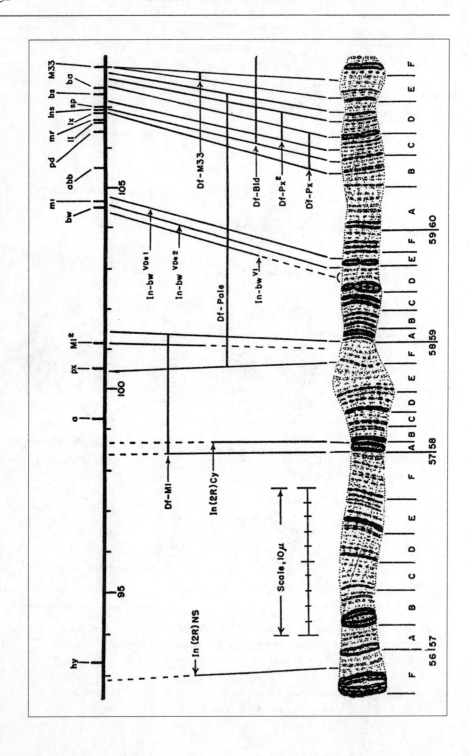

who was born in Kentucky. He studied an organism well known to all modern biologists, the fruit fly *Drosophila*. This little fly seems a peculiar choice for geneticists to study, but Morgan selected it in 1908 because of a curious set of characteristics which made it the perfect subject for research. It is remarkably prolific – in a warm environment *Drosophila* can pass from new-laid egg to sexual maturity in just 12 days. It is also a hardy species, and can withstand environmental circumstances that others might not survive. When reared in captivity, it reveals a number of easily recognizable mutations which scientists can study. Finally (and perhaps most important of all) we have the unusual nature of its chromosomes. The cell nuclei of *Drosophila* contain just four pairs of chromosomes, each differing in size and appearance and therefore easily recognizable.

Most remarkable of all, the chromosomes within the fly's salivary glands have giant chromosomes within them. Each genetic locus along the ribbon-like structure of the chromosome is multiplied so that the entire chromosome is enormously larger than normal. The result gives each chromosome a banded appearance, like discs on a string, and these discs can be identified with a specific genetic characteristic. It is almost as though you can see genes themselves using nothing more than a conventional microscope. This view took many decades to arise, however; the 1950 edition of Edmund Sinnott's book *Principles of Genetics* emphasizes that this idea had still to be proved, even at that relatively late date. Sinnott says that: 'a working hypothesis is that there exists a one-to-one correspondence between the genes and the ultimate discs in the salivary-gland chromosomes' but this was, he says, still 'an unproven hypothesis'.

Morgan had originally begun his own investigations when reports of Mendel's long-forgotten papers started to circulate. His original intention was to disprove Mendel's theories of inheritance, and at first it seemed as though his scepticism was justified. Morgan had observed that one of the mutations of *Drosophila* – a variety in which the eyes changed colour to white – did not follow Mendel's precepts. Instead, this characteristic

Opposite: **Mapping genes on fruit-fly chromosomes.** By studying the way genes crossed over from one chromosome, scientists had a way of finding where each gene was positioned. As the location data were gathered (across the top), they could be compared with the bands observed on the giant chromosomes. This map was published by C.B. Bridges in 1935. It shows part of the second chromosome of the fruit fly *Drosophila*.

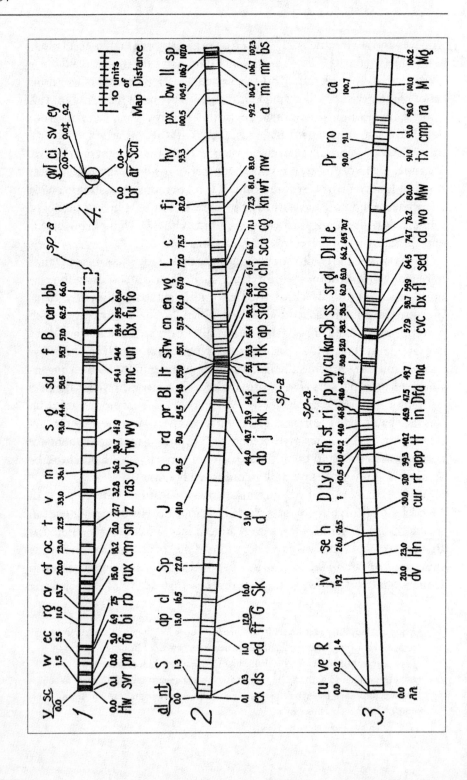

occurred almost exclusively in male flies, which (according to Mendelian inheritance) was an inexplicable phenomenon. After breeding several more generations of the fruit flies, Morgan realized that this could be explained perfectly – if the gene for white eye was found only on the chromosome that coded for the gender of the fly. He had discovered sex-linked inheritance.

He also noticed something that proved to be of crucial importance. The genes on a specific chromosome were not always inherited as predicted. Sometimes they were exchanged for genes on other chromosomes, and it was suddenly clear that chromosomes lying together during the early stages of cell division could swap genes with each other. This phenomenon of crossing-over was a revelation – and it provided Morgan and his team with a new research tool. They assumed (correctly, as it turned out) that the further apart two genes lay on a given chromosome, the greater the chance that one would be exchanged while the other was not. Genes that were closely adjacent would be statistically less likely to be exchanged independently.

For the first time, records of genetic exchanges could be used to calculate how far apart a specific pair of genes might be. One of Morgan's postgraduate students, A.H. Sturtevant, mapped five sex-linked genes using this statistical approach and produced the first-ever chromosome map. It was published in 1911. This set in train a massive research effort which eventually made *Drosophila* the best-known genetic subject in the world. By 1939 a total of 5,149 bands had been plotted on the chromosomes of *Drosophila*. Most people imagine that this kind of work has only happened in recent years.

It was clear that different organisms had different numbers of chromosomes, but the chromosomes of most animals proved very difficult to observe. They are smaller than the chromosomes of plants, as a rule, and counting them proved to be a demanding task. One of the crucial facts was the number of chromosomes in human cells, and estimates varied between

Opposite: **The first genome map takes shape.** During the 1930s the position of genes was systematically mapped on the four chromosomes of *Drosophila*. The genes were given short code names. This investigation set in train the genetic research that gave rise to the modern-day global project to map the human genome. We can now sequence genes and obtain accurate data more directly.

eight and sixty. Scientists were confined to counting chromosomes in the occasional cell in a specimen that was found in the act of dividing. In the normal interphase cell, between divisions, the chromosomes are not seen. Only when the cell divides do the chromosomes condense out of the nucleus and become visible. Even then, it is impossible to count the chromosomes unless the cell has been sectioned right across the central plane as it divides. This was a very rare occurrence, and hardly anybody observed human chromosomes clearly enough to count them properly. A microscopist at the University of Texas, Theophilus Painter, published the definitive answer in 1928: human cells contained 48 chromosomes, he reported. For 30 years this number was accepted, although he was wrong. The actual number is 46. They always say that Texans manage to make everything a little larger than reality, but that's not the reason for Painter's mistake. The chromosomes were such elusive structures, and were so rarely seen clearly, that his error is understandable. At the time, nobody was searching for a method of making chromosomes easier to study. It was believed that their elusiveness was a simple fact of life. Not until 1951 was the figure corrected. This final resolution was the result of a chance discovery by a research student at Galveston, T. C. Hsu. He was a PhD who had trained under Painter, and had been given the task of studying mammalian chromosomes. He found the task frustrating. Even when studying rapidly growing tissues, in which dividing cells were often seen, he was confused by the way the chromosomes were bunched together in the centre of the cell. One afternoon he placed a preparation under his microscope and was faced with chromosomes that were no longer confined within the cell. Instead, they were beautifully spread out. He was so surprised that he did not believe his eyes, and went for a walk around to the campus coffee shop. When he returned to the laboratory, he went again to look down his microscope and found that the chromosomes were still as he remembered. Through some fluke, they were scattered across the field of view and could be observed easily.

It turned out that the cells on this slide had been flooded with water, or a weaker saline solution than normal. The osmotic effects had burst the cells open, releasing their chromosomes and making them far easier to study. Hsu did not correct the mistaken notion that the human nucleus contined 48 chromosomes (he accepted this figure from his mentor, Theophilus Painter, and never doubted it). This accidental discovery was adopted by other chromosome specialists, who began flooding their cell

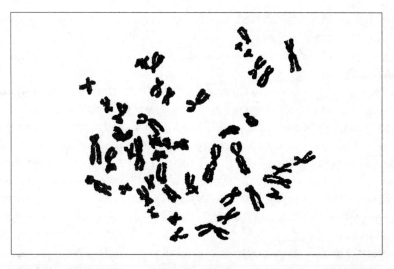

The chromosomes within each human cell. The 46 chromosomes that are typical of human cells can now be separated, stained and counted. Sometimes we find abnormalities. In Down's syndrome patients, for instance, there is an extra chromosome 21. In some traditional crop species we now know that the number of chromosomes is twice, or even four times, the normal value.

preparations with weak saline so that the cells burst open and released their chromosomes. It quickly became a popular technique.

Meanwhile Albert Levan, an American scientist, and Jo Hin Tijo in Sweden, made a further important discovery. Colchicine, an extract from the stamens of *Colchicum autumnale* (the autumn crocus), arrested cell division at the half-way stage. Colchicine prevents the formation of the spindle, which draws the chromosomes apart as the cell divides. As a result, cells start dividing but can never complete the process. A culture begins to collect cells arrested half-way through cell division. This meant that scientists were able to collect large numbers of dividing cells, and the search for occasional examples of cells in division was over. Levan and Tijo performed counts of the chromosomes in human cells and came up with the total of 46. In 1955 they announced their results, and this has been the accepted number of human chromosomes ever since. At least, this is the number in normal human cells. Jerome Lejeune, a French scientist, made a study of the chromosomes from cells of Down's syndrome patients. He determined that these patients had an extra small chromosome, and in 1959 announced that they had 47 chromosomes rather than 46. This was the first time that a congenital human condition was linked to an abnormality of the chromosomes.

Genes do not always stay in the same place on the chromosome – a remarkable and important fact. This phenomenon, sometimes known as the 'jumping gene', was discovered by an American scientist, Barbara McClintock, who worked on maize plants at the Cold Spring Habor laboratory on Long Island. She used the colours of the maize seeds, packed together on the cobs of corn, to study how inheritance worked. In time she plotted enough results to show how the genes were passed on from one generation to the next. Her accurate results also revealed something else: sometimes completely unexpected patterns of inheritance were observed. The only way of explaining the results, she felt, was to conclude that genes could change their positions along a chromosome. As her work went on, she narrowed her enquiries to two genes on the ninth chromosome: the Ac (activator) gene and the Ds (disassociation) gene that it controls. McClintock showed that a signal from the Ac gene actually makes the Ds gene jump to new positions along the chromosome. She could follow the movement by observing the effect that the Ds gene had on its neighbours. This upsets so many basic assumptions that it was clearly a most fundamental discovery. The findings were released at a meeting in 1951. I know what you are thinking: this came as a tremendous surprise to the geneticists gathered at the Cold Spring Harbor symposium that year, but this is not what happened. In the event, nobody took any notice of the work. It gave rise to little if any discussion, and passed without comment. McClintock retired to her laboratory and continued her work in private without further embarrassment. It took years before anyone realized what she had discovered. In time, the establishment recognized her importance, and indeed she was the recipient of the Nobel Prize for Physiology and Medicine – but not until 1983, more than 30 years after her results were announced.

Equally important was the way that genes can be passed from one cell to another. The observation was made by a British microbiologist, Frederick Griffith, in 1928. The target of his research was a vaccine against pneumonia. In the course of his research, he made an astonishing observation which seems surprising even today. He worked with two strains of bacteria, the S-strain being lethal, whereas the R-strain did not harm its host. Griffith had experimented on mice, which showed how the two strains differed so fundamentally in their effects on the host. His results showed clearly that the S-strain killed mice; the R-strain left them unaffected. What happened if the S-strain was killed by heat? As you would anticipate, the mice survived. However, if these mice were later injected with R-strain bacteria, they died. The living but harmless strain of bacteria was clearly capable of picking up the lethal gene from the dead S-strain organisms that they encountered. Later experiments showed that the same effect happened even in the test tube: a mixture of live R-strain and dead S-strain organisms gave rise to bacteria with the characteristics of live S-strain. The R-strain organisms had picked up something from the liquid surrounding the dead S-strain bacteria. The question was what?

For the answer to this riddle we cross the Atlantic to the Rockefeller University, where three Americans set out to track down the culprit in 1944. The team, led by Oswald Avery, repeated Griffith's experiments and then removed one component after another from the liquid, testing each time to see if the vital compound was still there. They removed the normal proteins and fats, then the carbohydrates, until they found that the infective principle remained when there was nothing left in the liquid but a strange and tenuous molecule. It had the consistency of sticky string. Avery found that he could wind it up on the end of a glass rod. They analysed the compound, knowing that it held the genetic clue to the transfer of infectivity, only to find that it was deoxyribonucleic acid or DNA.

Scientists needed to know how genes were made, and the race to unravel the structure of DNA was launched by Linus Pauling, an eccentric scientist at the California Institute of Technology. He has the remarkable distinction of winning two Nobel Prizes in different fields – the 1954 Prize for Chemistry (for his work on intermolecular forces) and the 1962 Peace Prize, commemorating his campaign against nuclear weapons. At the time, analytical methods from the field of chemistry were being brought to bear on the problem of analysing DNA. One of the most important techniques was X-ray crystallography, which went through many difficult periods before it emerged as a tried and tested technique. It is not difficult to understand. Look at a distant street lamp through an umbrella at night, and you will not see the lamp as a single image, but as a pattern of dots which covers a greater area than the image of the lamp. The light is diffracted by the threads in the covering of the umbrella, reflecting back from the individual strands in the weave, and creating a distinctive pattern. The pattern is directly related to the alignment of the threads – indeed, you could work out how the fabric was constructed by studying the pattern. Just as light can be diffracted by fabric, X-rays can be diffracted by the layers of molecules in a crystalline structure. The idea had first been proposed in 1912 by a German physicist, Max von Laue, at the Institute of Theoretical Physics in Munich.

When he discussed it with the Director of the Institute, Arnold Sommerfeld, he was met with severe disapproval. Sommerfeld could not see the point, and Laue was told not to waste any more time on the idea. He decided to defy his instructions, and gave the project to two young research students, W. Friedrich and P. Knipping. They soon produced the first X-ray diffraction patterns. At the time it was little more than a curiosity – indeed, Friedrich later admitted that they didn't really understand what was going on, and went off 'in an unfavourable direction' with their research.

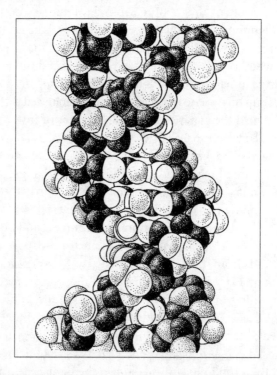

The familiar double helix of DNA. No molecule has ever been as recognizable as DNA. It is used to advertise events and exhibitions all around the world. It is a spiral structure which can separate easily into two, producing a new strand of DNA as it goes. New base-pairs are added to the side branches as the strands separate. There are three billion base-pairs in the DNA of a person.

It was two British scientists who were to spearhead the development of the technique into a standard laboratory procedure at Cambridge University. They were William Henry Bragg and his son Lawrence (who had also been christened William, but used his second name to avoid confusion). They were jointly awarded the Nobel Prize for Physics in 1915 for their pioneering work on X-ray diffraction. In 1937, Max Perutz, working at the laboratory established by the Braggs at Cambridge, set out to analyse the complicated structure of haemoglobin. It was a long task, and by 1947 his grant was running out. The University offered no further assistance. In desperation, Lawrence Bragg met the Secretary of the Medical Research Council for lunch at the Athenaeum Club in London, and managed to clinch further support in the nick of time. In the end, the structure of haemoglobin was worked out through this technique, and Max Perutz was jointly awarded the 1962 Nobel Prize for Chemistry. The other award winner was his co-worker John Kendrew, who had used X-ray crystallography to work out the structure of myoglobin, a muscle protein.

During the subsequent decades it was shown that mutations could be caused by cancer-causing chemicals and radiation. Many workers identified a chemical called DNA in the chromosomes, and in the 1940s the idea began to crystallize that this DNA held the clue to the nature of the genes. The problem was to find a chemical structure that revealed how DNA could be reproduced, and handed on down the generations. The subject received a tremendous boost when it was raised in a major meeting – but the meeting only took place because of a furious row. At the end of World War II, the Society for Experimental Biology decided to celebrate with a grand conference on 'respiration'. A serious quarrel developed over the organization, and at the last moment it was decided to change the subject and avoid further disputes. One of the research team suggested 'nucleic acids' as an alternative topic that would irritate no-one, and this was agreed. It is strange to realize that this crucial meeting of July 1946 at Cambridge owed its existence to a display of temper. In the scramble to find enough papers, some research from years earlier was dug out and presented. It included an X-ray diffraction pattern for DNA. This served to rekindle interest and, from this time onwards, it was accepted that DNA was the genetic material of the cell, and the race was on to unravel how it could be replicated as a cell divides.

Two important scientists figure in the research on the replication of DNA and no, I am not thinking of Crick and Watson. The two who produced the evidence for the strange nature of the DNA molecule – the celebrated helix, as everyone now knows – were named Wilkins and Franklin. Maurice Wilkins, born in 1916 in New Zealand, was fascinated by the study of the key proteins within the nucleus. In 1951 he gave a lecture which emphasized: 'The study of crystalline nucleoproteins in living cells may help one approach more closely the problem of gene structure.' Wilkins recognized that the DNA molecule was shaped like a spiral, and the idea of a helix was born. Rosalind Franklin was a brilliant young crystallographer, and it was she who made the crucial discoveries about the structure of DNA which led people to the final version – the double helix. This answered the problem of how genes reproduced. The spiral slowly unravelled, unwinding like untangling the cord of a telephone, producing a new spiral of DNA as it went.

Francis Crick (born in Northampton in 1919) and James Watson (born in Chicago in 1928) were great opportunists. Crick was a physics graduate who worked on magnetic mines before becoming fascinated by biology. By 1949 he had joined the Medical Research Council laboratories in Cambridge. Watson was a precocious child. He was originally 'discovered' by Louis Cowan, producer of the Chicago Kid Quiz Show. At that time, Chicago University introduced an experimental scheme to admit youngsters to its degrees, and Watson was taken on at the age of 15. Although he studied embryology, his main interest remained bird watching and he enrolled for an ornithology summer course in 1947 at the University of Michigan. He even wanted to specialize in the subject when he applied to Indiana University as a graduate student. Crick and Watson met at Cambridge. They resolved to work on solving the way that molecules of DNA could divide, and consulted widely with those who had been working on the problem. Inevitably, this caused problems. Franklin was angry that her crucial results were being used by people who did not fully understand the problem. The brooding discontent has been vividly portrayed by the eminent historian of science, Robert Olby, who is dismissive of a proposal by Watson and Crick that they might join forces with Rosalind Franklin and her colleagues for a joint project on DNA. Olby says that Franklin and her team would have nothing to do with the suggestion. 'They had witnessed two clowns up to pranks,' says Olby. 'Why should they condone their behaviour by joining forces with them?' After discussions with leading research scientists in the field, Lawrence Bragg decided to place a ban on further research in this area by Crick and Watson. The field would be left to the specialists. Crick was given a project on 'X-ray studies of polypeptides and proteins', and Watson was told to work on tobacco mosaic virus.

An eminent chemist, Erwin Chargaff, had been using the analytical technique of chromatography to look at the chemical structure of DNA and had discovered an astonishing fact – for all the complexity of the information that DNA contains, it has only four different bases. It is like finding an alphabet with just four characters instead of the 26 in English. Bear in mind that English uses different combinations of just 26 symbols to express our breadth of knowledge and the great works of master authors. However, the number of letters in an alphabet is not related to the quality of the literature it spawns. Other alphabets have greatly differing complexity. There are 65 other alphabets in use around the world. The longest is Khmer which is spoken by the Cambodian people and comprises 74 characters. The shortest is Rotokas, spoken on Bougainville Island, Papua New Guinea, made up of just 11 letters. Well, the alphabet of DNA is shorter even than that, with just four. Using these in different combinations, and in varying sequences, nature contrives to build entire humans from simple chemical compounds. One cannot begin to comprehend such awe-inspiring complexity. In the spring of 1952, Chargaff met with Crick and Watson for a discussion about the structure of DNA. He censored them for knowing little of his own major discoveries, and dismissed the pair as 'pitchmen' (which, Olby explains, meant 'salesmen or ad-men shooting a line'). Chargaff said: 'I have never met two men who knew so little and aspired to so much.'

There had been several possible models for DNA proposed by different teams, but none was entirely satisfactory. In 1952, Crick consulted a noted Cambridge mathematician, John Griffith, on the possible combinations of adenine or thymine and cytosine or guanine which could explain the structure of DNA. Crick and Watson persisted in trying to build their own model for the construction of DNA, but could still not find one to fit all the facts. One possible solution was an elegant structure proposed by Watson late in February 1953, but on 27 February Crick pointed out that there were still two remaining problems, around which it was difficult to fudge the figures. The idea collapsed, and they departed to think up an answer. Next day, Watson got out the model kit again and began to reorganize the bases – and this time it seemed to work. He wrote 'all the hydrogen bonds seemed to form naturally; no fudging was necessary'. When Crick came in a little later, he agreed with the answer Watson was proposing and the model of DNA seemed successfully within reach.

Crick strode along Free School Lane for a beer at The Eagle. He walked into the bar telling everyone that they had just found the 'secret of life'. Wilkins wrote them a letter saying, 'I think you are a couple of old rogues', but agreeing that their model did seem to fit the known facts. Bragg, who had tried to stop their research, said: 'Well, it's all Greek to me.' In the edition of *Nature* published on 25 April 1953, the model was published in a short paper by Crick and Watson. It was followed by one from Wilkins and his team, and after that came a third paper by Rosalind Franklin. Crick, Watson and Wilkins shared the 1962 Nobel Prize for Medicine; Franklin had died of leukaemia by the time she was 30.

The sequence of amino acids in DNA was worked out by British biochemist Fred Sanger, born in 1918 in Gloucestershire. He was studying the sequence of amino acids in proteins. In 1943, Sanger began to unravel the sequence of amino acids in the insulin molecule, and in 1945 he discovered that he could cut chains of amino acids into lengths using 2,4-dinitrofluorobenzene. It took him 10 years to unravel the structure of this important substance. By the time his project was complete, in 1953, he was able to announce that there were small differences between the forms of insulin found in different species. He next turned to sequencing RNA and DNA, and several different enzymes were discovered around the world which could cut DNA into lengths and recombine them. This gave birth to the recombinant technologies that are widely used to unravel the structure of genes. One particularly useful tool is a small circle of DNA which can be removed from cells, copied in large numbers, altered and then put back in again. These circles are known as plasmids, and are an important tool used by geneticists.

Then came the crucial discovery of PCR by Kary Mullis (b. 1944). He had been working in a restaurant when he was tempted back to join the Cetus Corporation of California. He found that you could take DNA and heat it with polymerase extracted from a bacterium found in hot springs, and thus unaffected by the heat of the experiment. The resulting polymerase chain reaction (PCR) triggered the DNA to keep reproducing itself to provide huge numbers of identical copies. This gave us a chance to recover DNA from tiny samples, and gave rise to the idea of Jurassic Park.

Do not imagine that the scientific world was immediately set alight by his ideas. His papers were rejected by the major scientific journals, and his work was looked at without interest by the scientists at Cetus. Eventually they paid him a bonus of just $10,000, while Roche bought out the PCR patents from Cetus for $300,000,000. Mullis went on to win a Nobel Prize for his discovery in 1993. Now he has a private institute in California. Each evening he roller-blades, and he goes surfing at dawn. Meanwhile, papers on PCR are published at the rate of 100 a day.

Analysis of DNA has made it possible to trace the relationships between families of plants and animals, to provide evidence in murder trials and to unravel a family's history. We will soon be using DNA testing to control the quality of food (and even to look for orange juice that is not as pure as it claims). Scientists in the far reaches of the Arctic tundra are hoping to retrieve sperm samples from frozen mammoths, which offers the possibility that we may revive these fabulous creatures.

It took three centuries to move through this complex tapestry of intrigue and rivalry, personal curiosity and single-minded pushiness, from the first recognition of the cell to the harnessing of recombinant technology. From now on the pace of progress will be unimaginably increased. Physicists are starting to show an interest in DNA. They can twist the molecule into unnatural shapes and harness its unique properties. They have unzipped the double helix into its two halves, measuring the force that normally holds them together. Scientists from several different disciplines have hopes of using DNA as a microscopic string to help them solve problems that have nothing to do with genetics. Everyone needs to keep in touch with the research. Only then can we hope to understand what is happening – and only if we understand it can we hope to control the future.

3

How cells began

IT IS THE LIVING cell that has made our world the way it is. Single cells have had an enormous effect on our environment and on human society. Much of our landscape has been produced by the action of microbes. The white cliffs of Dover and the limestone Cotswolds were built by microbe communities that lived millions of years ago. During their lives, they laid down tiny mineral shells which accumulated in droves at the bottom of the seas. Over millions of years these strata hardened into rocky layers which were thrust upwards by geological action and weathered to produce the landscape of today. Elsewhere in the world, different types of organisms have bequeathed their remains to the modern world. At Lompoc in California there are huge deposits of a white and crumbly rock. It is made of finely divided glass. The rocky cliffs are composed of the fossil skeletons of single-celled diatoms. These beautiful little algae strengthen their cells by constructing an internal skeleton of pure silica and, when the cells die in huge drifts, the glassy structures remain. Where there have been large deposits the resulting mineral (diatomite) is mined. Diatomite has been used in toothpaste (where it acts as a mild abrasive) and in the manufacture of dynamite (where it absorbs, and stabilizes, nitroglycerine).

Living microbes are regulating our environment today. When the leaves fall in autumn, park keepers go round to sweep them up and dispose of them. Nobody sweeps up the leaves that fall from trees in a nearby field, yet within a week or two they also disappear. Where do they go? It is the teeming populations of microbes in the soil that do the work, breaking down the leaves and returning their essential goodness to the soil where they are

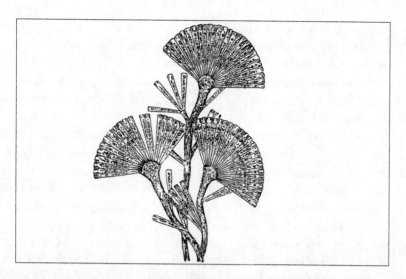

Pond diatoms with shells of pure glass. The diatoms are algae which produce an inner skeleton of silica to support the cell. They secrete oil, and today's oil fields probably result from vast communities of diatoms. Huge cliffs of pure diatom shells (known as diatomite) are still mined at Lompoc, California, and are used to make dynamite and toothpaste. These delicate illustrations were made by William Carpenter in the 1850s.

utilized by a future generation of plants. Apart from the flattened remains of animals killed on the highway, you rarely see the dead body of an animal out in the wilds. Within days of dying, an animal carcass has been colonized by microbes. In a co-ordinated sequence they break down the remains to soluble components which can be digested and returned to the soil. The world would be in an uninhabitable and disgusting state without the ceaseless activities of microbes.

It is to the microbe world that we owe our breathable atmosphere. Surprising as it seems, the trees in our forests do very little to provide us with oxygen. It is true that, as a green plant grows in the sunlight, it gives out oxygen. However, plants do not do this all the time. During the hours of darkness they consume oxygen and give out carbon dioxide, just as we do, for then they are metabolizing their food reserves and using them as an energy source. Once a tree has fully grown, it has made a net contribution of oxygen to the air. That, however, is only half the story. As the tree dies and decomposes, it is broken down by the microbes of decay. During this process the tree is recycled so that its components become available to other organisms in the environment, and the oxygen it gave out during its life is consumed as it decays. If the tree is burned as fuel, then the oxygen consumed as the flames burn brightly equals the oxygen given out as the wood was formed. The ash that remains is composed of the chemical constituents taken up by the tree as it grew, but almost all of its bulk was made up of carbon dioxide and water.

Photosynthesis is the process by which a plant captures the energy of sunlight. It is coded for by the genes within plant cells, and (although we understand the process well enough) it is something that science cannot imitate. Photosynthesis removes the oxygen from carbon dioxide molecules and releases it back into the air. The atoms of carbon that remain are chemically bonded to the water, producing hydrocarbons and other compounds as the tree grows. The process is powered by sunlight, which the green chlorophyll captures every second that the sun is shining on the leaves. As the wood burns, the process is reversed. The carbon is broken free from the elements of water and joined with oxygen from the air (this is why a fire goes out if there is not enough oxygen). This re-forms carbon dioxide. The elements of water are given off as steam, and the energy that shone from the sun as the plant grew is released as heat. Thus, at the end of the cycle, the tree has been restored to its original components of carbon dioxide and water, the heat energy absorbed from the sun is given out in the fire, and the oxygen originally released is reabsorbed.

So where does the oxygen come from? Mainly from microbes. The surface layers of the oceans are rich in a soup of tiny green algae. These grow and reproduce in the sunshine, releasing oxygen into the atmosphere and surrounding sea water. Masses of senescent cells drift down, forming a deep layer on the ocean bed, and it is in these that the captured carbon resides. The oxygen released (as the carbon dioxide was split into oxygen and carbon) remains in the atmosphere, and will remain there unless the deep sea oozes from the oceans could be somehow brought to the surface and completely decomposed.

Although one of the benefits of an atmosphere of oxygen is the phenomenon of human life, oxygen is a dangerous gas. It is a corrosive substance. It oxidizes anything it can. The energy trapped in objects around us is vast, and it takes a small flame to set off a train of widespread oxidation as the oxygen in the air wreaks its havoc. Fires – forest fires, the Great Fires of Chicago, London and San Francisco, fires that wipe out families and decimate a landscape – are among the most feared of all threats. Oxygen is the cause of these disasters. The standard test for oxygen utilizes a glowing splint: take a thin sliver of wood, set fire to the tip so that it burns like an elongated match, and then blow out the flame. The faintly glowing tip of the splint, if plunged into a gas jar of oxygen, bursts immediately into dazzling white flame. That's why the first instruction in an aircraft is to extinguish all smoking materials before using an oxygen mask. Disregard that elementary instruction and you risk being enveloped in flame. It is the energy liberated by oxygen when it attacks the carbohydrates in our bloodstream that provides the source of warmth for our bodies. So energetic is the chemical reaction that the heat it gives out warms us up to 37°C. It would go higher without control (just as it does in a fever) so we have a complex system of cooling mechanisms that prevent our getting too hot. Sometimes a general anaesthetic can trigger a loss of temperature regulation, and malignant hyperthermia results. There is little the doctor can do, and the patient's uncontrolled use of oxygen in the body causes a fatal rise in temperature. The patient cooks to death.

So great is the energy stored in the body that, in some rare but well-documented instances, an individual going about his or her daily affairs bursts into flame. This phenomenon of spontaneous combustion has featured in novels, in folklore and also in the scientific literature. It is testimony to the power of oxygen when its action runs out of control. An early example was that of an elderly woman in seventeenth-century Essex who was found burnt to death in her bed. There was no sign of combustion anywhere else in the house, and even the bed on which she had been lying was relatively undamaged. A contemporary commentator wrote: 'No man knoweth what this doth portend', and the cause of the phenomenon has remained a mystery to this day. More recently, a building contractor in Yorkshire drove past one of his sites, waved to the workers through the car window, and was immediately immolated in a ball of fire. The rest of the car was relatively unscathed. In London, the *Daily Telegraph* reported the case of a truck driver found incinerated in the cab. The inquest found that the interior of the cab was badly scorched, the driver himself little more than ashes, yet the fuel tank was intact and the rest of the vehicle unmarked. In a now-defunct evening newspaper published in London, *Reynold's News*, a reporter described the case of a west London man who burst into flame in the street. 'He appeared to explode,' said the account. 'His clothes burnt fiercely, his hair was burnt off, and the rubber-soled boots melted onto his feet.' In the late 1950s a young dancer at a discothèque suddenly burst into flames. At the inquest, her boyfriend (his account corroborated by other witnesses) explained: 'I saw no-one smoking on the dance floor. There were no candles on the tables and I did not see her dress catch fire from anything. I know it sounds incredible, but it appeared to me that the flames burst outwards, as if they originated within her body.' The Coroner recorded a verdict of death by misadventure, adding as a unique codicil: 'Caused by a fire of unknown origin.'

The energy stored within a single person would be quite enough to support a conflagration, but there are clear problems in understanding how spontaneous combustion could occur. Exactly how burning might be initiated is one fact that we need to explain, and the watery environment within the living body also militates against the phenomenon. Answers to the first of those points have been proposed in the past, and I will offer a solution to the second problem. The consensus seems to be that static electricity could provide the spark that ignites the fire. Robin Beach, a consultant of Brooklyn, set up a series of tests to see how much electrical charge a person might normally acquire. By standing a person on an insulated metal plate he could measure any charge by means of an electrostatic voltmeter. The highest reading he obtained was from a young female employee who recorded 30,000 volts, so he recommended that she be transferred to a part of the company well away from inflammable materials. Beach discovered that readings in excess of 10,000 volts were found in people who had dry skin and who (in low-humidity conditions) had walked across carpet. It has long been known that static electricity results from this kind of situation; indeed, an early method of demonstrating a static charge was to rub an ebonite rod with a silk handkerchief. Many of us will have seen sparks fly if someone removes clothing in a dark room (notably if the air is dry and the garment is made of an artificial fibre), and you can even hear static sparks fly when long hair is brushed. The estimate by Beach was that perhaps one person in 100,000 has uncharacteristically dry skin, so a static charge adjacent to the body surface is less likely to be discharged through a film of moisture. He thought that people bursting into flame caused damage to American industry worth tens of millions of dollars each year.

The essential question he could not answer was this: how could a tiny spark become translated into a lethal conflagration? I think the answer lies in an error of human metabolism which some people manifest, particularly when ill. I refer to the formation of acetone. Many illnesses cause a subtle change in the body chemistry, in which the body starts producing acetone as a metabolic by-product. You may have smelt acetone on the breath of a sick child (a baby sometimes manifests this when teething, for instance). One important consequence of this is a groundless accusation that a child has been indulging in glue-sniffing. The advisory literature circulated to anxious parents offers the detection of a solvent odour on the breath as a tell-tale sign of solvent abuse. Only rarely is this the case. Most children exhale acetone as a result of a mild illness, and not because they inhaled it. Two characteristics of acetone are of the greatest interest to me in this context:

1. First, acetone is highly inflammable; indeed a mixture of acetone and air is violently explosive.
2. Second, acetone is an excellent fat solvent. When produced in the body, acetone could saturate the fatty layers.

We would thus have a person with a layer of (combustible) energy-rich fats saturated with a (highly inflammable) solvent. When you bear in mind that the blood supply is enriching the body with oxygen, so vital for combustion to start, I believe that we could have a situation in which the body was primed, ready to ignite. It seems to me that this offers a possible explanation for one of the most perplexing of tragedies. Meanwhile, it reminds us of the power of the body as an energy store, and also makes one realize that oxygen is a potential hazard unless its chemical reactions are properly regulated.

Meanwhile, look what happens when a solution of oxygen gets at your car. We make steel by blasting iron oxide with heat energy, and drawing away the oxygen by combining it with carbon. The result is that we end up with iron, and the carbon dioxide by-product billows out into the air. Iron (which, when it contains a little dissolved carbon, changes its name to steel) is then processed to make motor cars. From that moment on, the process begins to reverse and we start to move back again to iron oxide. As water and oxygen get at the body of your car, the iron oxide begins to reappear. We call it rust. Leave it untouched for long enough, and the car will break down to form a brown heap of iron oxide. We use 'steel' as a metaphor for strength, yet water and oxygen can corrode it to dust. Let nobody believe that oxygen is somehow harmless. It is a powerfully reactive chemical, and the energy of its reactions is what powers planes to travel faster than sound, and powers the energetic people who design them.

When the earth was formed, free oxygen must have been a rarity. So reactive is this element, that it can only have been present in a form combined with others: the compounds called oxides. One of the most important of these is an oxide of hydrogen, H_2O – water. Hydrogen burns fiercely. It is a hungry atom, longing for chemical partners. It is also the simplest atom of all, with a single proton at its core balanced by a lone electron orbiting around it. In the early earth, there were many compounds of hydrogen. Water is an amazing molecule, and water is the only known substance that exists in all three phases – solid, liquid and gas – on the earth. It is also one of the very few substances that expands when it solidifies. This is important for life: if water froze from the bottom up, any organisms in the water would be brought to the surface and die. It is because water does not obey the normal laws of science that life is possible.

Apart from H_2O (water), there was also H_4C (methane, conventionally written the other way about, CH_4), H_3N (ammonia, written as NH_3), and H_2S (hydrogen sulphide). Hydrogen sulphide is a poisonous gas to us, with strict exposure limits imposed in the workplace, but plenty of microbes can utilize it as an energy source. Some produce it too, which is why a bout of intestinal upset can cause you to produce hydrogen sulphide. As the gas is rich in energy (and therefore inflammable), it is always best not to stand with your back to an open fire if you are likely to produce an invisible cloud of hydrogen sulphide after a bad meal. This is the gas produced by the microbes that turn eggs rotten, of course, and it is potentially explosive. Try explaining that to a junior doctor in the emergency clinic.

Hydrogen forms a range of acids, combining with chlorine to form HCl (hydrochloric acid), adding oxygen to hydrogen sulphide to form H_2SO_4 (sulphuric acid), and combining with nitrogen and oxygen to give HNO_3 (nitric acid). A single atom each of hydrogen, carbon and nitrogen forms an acid we regard as a deadly poison, HCN (hydrogen cyanide). Yet if you line up five of those together and join them chemically, you form $H_5C_5N_5$ and this is known as adenine (see the figure on page 155). Adenine is a fundamental part of the adenosine triphosphate (ATP) molecule, the seat of energy release in the living cell; it is also a key component of DNA itself. If we look at the planets orbiting with us around the sun, we find huge amounts of these compounds. When we look at the chemistry of life on earth, we find derivatives of these compounds everywhere and in every living cell. A watery solution of proteins makes up the bulk of the cell contents or cytoplasm, and proteins are the stuff of life. Should you wish to have first-hand experience of what the contents of a living cell are like, then I can direct you to the largest cell you normally encounter in your daily lives. It is a hen's egg. The unfertilized egg of a chicken is a single cell. Tucked away on the edge of the yolk is the microscopic cell nucleus in which the genes are found, but the white of the egg is pure cytoplasm. Each cell of your body is filled with something very similar. It is translucent and wet, slimy and soft, and composed of a matrix of proteins dissolved in water. Many cells (particularly single cells which have to exist independently) possess storage bodies within them, and the yolk of the egg is a typical example of stored food reserves inside a living cell. For all the complexity of the protein system, proteins are made of amino acids – and amino acids are molecules that are made from carbon, nitrogen and water. Our innermost nature embodies resonances of the chemical composition of the primitive earth, when earliest life began. The move from a hot world of bubbling lava flows and exploding volcanoes to a more peaceful planet populated with people centres on a single crucial episode – the creation of life itself.

Even our own biochemistry reminds us that the most complex forms of life contain the same chemical components as that early earth. What was needed were mechanisms to combine those chemicals so that they could start to reproduce themselves. There have been many experiments in laboratories which may throw some light on the problem, and the most famous of those were carried out in 1953 in the laboratories of Harold Urey at the University of Chicago. The approach was very simple. It has long been known that plants derive much of their nitrogenous nutriment from the air – lightning flashes provide so much energy that the inert nitrogen in the atmosphere forms nitrates which descend in rainfall. What else might lightning produce? One of Urey's graduate students, Stanley Miller, took a selection of the simple molecules that we have already encountered – water and ammonia, hydrogen and methane – and enclosed them in a glass flask. The flask was fitted with electrodes as a supply of energy, and he set up an electrical discharge across them. In this way, he had compounds from an early earth energized by captive lightning in the laboratory. He left it like this for a week, and then analysed the results.

In the mixture that resulted, they found a range of complex chemical compounds. Most satisfying of all, they discovered alanine and glycine – two amino acids of crucial importance to living organisms today. Since then, other research teams have managed to produce a great range of the same amino acids that make up the proteins in our bodies. Among the other crucial compounds that have been synthesized in this way is ATP which, as we have seen, is the fundamental energy store within all cells, and the trigger for the activity of life. More recently, Leslie Orgel at the Salk Institute in California managed to produce a long molecule containing 50 nucleotides somewhat reminiscent of the structure of DNA.

There is the problem of how these compounds could survive; in a primitive earth environment there are as many forces breaking such molecules down as there are building them up. At Cardiff in 1972, I began work on a new possibility: that life had actually originated on earth, but the molecules that gave rise to it had themselves been elaborated in space. We now know that you can find evidence of plenty of organic molecules in space, so there are reasons to believe that space chemistry could have provided the raw materials, and we have an abundant supply of radiant energy from the countless stars across the universe. Complex compounds that were formed in space would not be so liable to be broken down as those formed on the earth's surface. The fate of many complexes in a watery earth environment is that they become hydrolysed; in space this process of hydrolysis would be much reduced. By 1973, the press had reported the modest progress I had made, and the idea has remained popular at Cardiff ever since. I still believe that some of the earliest compounds that came before life might have been formed in space. If they were, then some of the pre-biotic molecules could have come to earth ready-made and ready for action.

Other research workers claim that life simply travelled to earth from outer space, spreading across the universe in comets or dust particles. For many decades we have known that there are occasional traces of organic molecules in some meteorites, and this has given rise to the so-called panspermia theory, one consequence of which suggests that life has been spread throughout the universe. It has preoccupied philosophers in many lands. Well, it does not preoccupy me. The panspermia idea does nothing to answer the question. Although it suggests that life could have originated a long way away, it does not begin to answer the question of how life actually began. An analysis of what happens in the chemistry of space, coupled with the magnificent results from laboratories that have produced organic compounds from the simplest raw materials, can one day allow us to model how early microbes might have originated.

Even chemical solutions can sometimes be shown to form little round bodies that look like primitive cells, and some of these have been shown to divide in half as they increase in size. This is not life, of course, but it does show that the way cells behave is mirrored by the properties of non-living complexes. We can see that a crude idea of the origin of life is within our comprehension.

The violence of the early earth cannot be imagined. Until four billion years ago there was not even a crust on the surface of the molten planet. Even now the crust moves about, with tectonic plates jostling against each other. The velocity with which they move is within our comprehension: the continents move at about the same speed as our fingernails grow. Some clue as to the conditions in an early ocean can perhaps be gained by studying the vents on the ocean floor, from which volcanic heating sends up jets of boiling water. There are strange, primitive life forms clustering around these vents; they take us back in time to an earlier period in our planet's development. The very first life forms, whatever they were, appeared surprisingly quickly after the earth cooled and its crust appeared. We believe that the crust started to form no earlier than four billion years ago; recent discoveries suggest that fossil microbes exist in rocks that were formed three and a half billion years ago. If this is the case, then the planet went from a turbulent and molten state to offering a life-supporting environment in less than 500 million years. There are even suggestions that the period may have been as short as 100 million years, which does give one reason to speculate on the likelihood of life having arisen elsewhere in the universe.

The earliest living organisms had to contend with compounds including hydrochloric acid and hydrogen sulphide. The forms of life with which we contend are not so very different. There is enough hydrochloric acid in your stomach to burn a hole in the carpet, for example; and although we find hydrogen sulphide (H_2S) a poisonous and offensive gas, if we exchange that sulphur atom for one of oxygen we end up with water (H_2O), which is essential to all forms of life. Strange as it seems, the earliest living organisms were not enough to lead on to the forms of life we see today. They gained the energy for living by breaking down chemical molecules and capturing the energy that was released. There is a limit to this: once the energy is gone, there is nothing more to sustain life. What was required was a means – not merely of releasing energy, but of capturing new energy. This is the role of photosynthesis, the harnessing of energy from the sun. In the modern world we think of green plants as the principal players in this arena, but the earliest photosynthetic organisms did not use the familiar chlorophyll. Their photosynthetic pigments were purple and red. With these organisms we are on more familiar ground, for they still exist today. Indeed, there are some forms of photosynthetic bacteria today that are virtually indistinguishable from fossil forms found in rocks more than two billion years old. We are harnessing the energy from those long-dead microbes throughout the modern world. Just as the hen's egg contains a reserve of yolk as an energy and nutrient reserve, single photosynthetic cells lay down stores of energy too. Some of them combine water and carbon dioxide to form hydrocarbons, others go further and store away droplets of oil inside the cell. Just as the great drifts of dying microbe cells laid down beds of limestone, chalk and diatomite, the vast growths of these organisms produced huge accumulations of oils – which we now extract through the oil wells we need to fuel our modern world. If you look at a single diatom cell under a microscope, you can see its glassy inner skeleton, and the refractive droplets of oil that it lays down for dark days ahead. Magnify that quadrillions of times, and you are laying down the oil fields on which human society now depends. The energy of the sun which those single cells captured is released in the furnace, and the warmth of a modern centrally heated building is the energy of the sun from a billion years ago, trapped by a microbe and put in store to fuel our future.

The variety of cells is coded by their genes, and the range of living organisms is truly stupendous. The most reckless invention of the science fiction artist cannot begin to match the astonishing variety of the cell in the real world. Some are so tiny that we need an electron microscope to make out their details, whereas others are so large that you can handle them with impunity. The bulkiest animal cell is, of course, a bird's egg and that (in the case of the ostrich) can weigh over 2 kilograms (4–5 pounds) and measure more than 20 centimetres (8 inches) from end to end. The largest plant cell must be that of a group of marine algae known as the Cauperlaceae, single cells of which can be up to a metre (three feet) in length. The most highly developed single-celled organisms are stunningly complex. Some contain eye spots that can clearly focus light on a retinal spot. Many species have contractile stems which allow them to float away in the surrounding water, but pluck them back to safety if attacked. Single-celled microbes have an impressive variety of systems for moving about. The simplest seem to be the amoebae, which flow along by projecting a new extension from the front of the cell while drawing in cell substance from the rear as they go. However, amoebae have immensely complex systems within the cell to enable them to perform this trick. For a formless blob of watery jelly to move about and feed is actually one of the hardest phenomena to understand. Other microbes have long and graceful flagella, which lash them along through the water, and many are covered with a field of beating cilia, which propel them along at a speed higher (size-for-size) than that of an Olympic swimmer. A few odd organisms have a single flagellum projecting from the front of the cell, only the tip of which is in motion. It draws the cell along, like a tiny mouse taking a bear for a walk on a leash.

We have a good idea of the way these complex cells arose from the simpler bacteria that were their ancestors. It seems that cells began to live within each other. The flagella, for example, with which some single-celled organisms swim are really captured, free-swimming bacteria, which existed long ago. The mitochondria inside the present-day cell (where they undertake the essential cell chemistry that keeps it alive) may have once been independent bacteria which took to living inside other cells through expediency. One reason for this belief is the presence of DNA inside mitochondria, which is a clear indication that they might have been self-contained microorganisms at some stage. In much the same way, the green bodies within plant cells that capture sunlight were once independent algae, living a life of their own before they became incorporated into a larger and more complex cell. This is a stunning idea – and it is one that is over a century old. The proposal that cells merged together in this way was proposed in 1893 by a Swiss scientist, Andreas Schimper, author of a tremendous book on global vegetation. He proposed that the little green bodies containing chlorophyll in higher plants had once been independent algae. By 1910, Konstantin Mereschovsky in Moscow had extended the theory to include bacteria incorporating themselves into cells to produce organelles like the flagella, with which some cells can swim, and a host of other structures besides.

There have been many experiments since that substantiate this revolutionary idea. Analysis of the DNA within some of these little bodies shows strong similarities to the DNA of primitive bacteria. Even more convincing are the observations of Kwang Jeol of the University of Tennessee. A culture of amoebae in his laboratory became infected with bacteria which threatened to wipe them out – but within weeks a few of the surviving amoebae were growing again with the bacteria living harmlessly inside them. The amoebae had changed to accommodate the bacteria, and the bacteria had altered their ways in order to live inside the amoebae without causing

Algae can show how multicellular life began. The fresh-water alga *Batrachospermum* produces jelly-covered strands reminiscent of the appearance of toad-spawn (the literal meaning of its name). A central strand of elongated cells forms a primitive stem, and radiating bunches of cells produce spherical globules of tissue strung out along the main axis. Reproductive bodies form inside the globules of cells.

problems. A completely new organism had been produced, and it had happened quite by chance. If the acquisition of new components happens so easily in nature, then new species might often have been produced by the merging of two (previously independent) lines of cells.

Such complex cells have a distinct nucleus, and their more specialized structure gives them tremendous advantages over the simpler purple bacteria and blue–green algae that had gone before. They first appeared about two billion years ago, by which time the atmosphere was already rich in the oxygen produced by the photosynthetic bacteria and blue–green algae which had excreted oxygen as they harnessed the energy of the sun. These earlier organisms, bereft of a nucleus, are now known as prokaryocytes; the more highly developed cells are classified as eukaryocytes. It is from the eukaryocytes that the multicellular organisms such as ourselves have descended. Cells of different types began to live in communities which functioned as a single organism about a billion years ago, and by 500 million years ago animals and plants were ready to leave the seas to colonize the land.

There remains a conceptual difficulty which bedevils current discussion. How can a single cell – destined to develop into a multicellular body

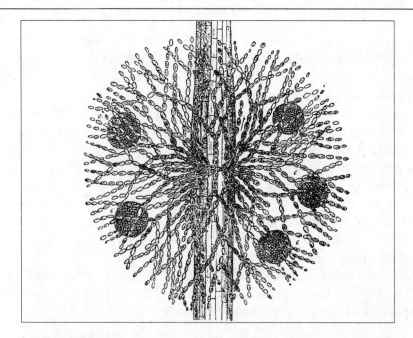

The toad-spawn alga under a high-power microscope. Looked at under higher
magnification, we can see how some cells specialize as the building blocks of the
central stem, others radiating out in carefully spaced arrays so that they do not
compete for light and dissolved gases. Cystocarps, the sexual organs, can be seen
within the cell mass. This shows the beginnings of development towards the vascular plants.

(like a person) – be a simple, single cell, and yet at the same time also
contain the elements of a multicellular organism? How do cells acquire the
very different shapes and forms which manifest themselves as a single
organism? The biologist Paul Weiss set out the problem in these words:

In the developed system, 'organism', the cells represent the elements ... but the pri-
mordium of the organism – the egg – does not consist of cells. Now, there arises a
dilemma.

I believe that the nature of this complicated process lies in the suppres-
sion of some facets of a cell's behaviour, and the emphasis of others.
According to this view, every cell begins with the potential for a full range
of propensities – movement and sensation, digestion and storage – but in
different regions of the body, and in different sites within the organs,
specific abilities become highly developed at the expense of others. The

cells of the retina would become specifically devoted to the task of light reception, for example, while eschewing movement. Those of the voluntary muscles which produce the movement of our limbs are specialized to respond only to nerve impulses and to contract in response, while losing their other facilities. Cells repress the functions they do not require, and allow full expression to those that apply to their specific function in the body. I believe that this is true whether we are speaking of multicellular animals or plants, small or large.

Today we know of these teeming populations of plants and animals of all shapes and sizes. Some exist as amazingly intricate single cells, whereas others take advantage of the cell's abilities to specialize and give different cells specific jobs to do. To me the key fact is that every plant and each individual animal remains a community of separate cells, and the way they behave is a manifestation of the behaviour of the separate cells comprising them. If we can come to terms with the lives of cells, we will truly begin to understand the origins of human nature.

4

Harnessing the cell

F ROM THE EARLIEST TIMES, humans have harnessed the energies of microbes to help advance their aims. The exploitation of
microbes is always based on the same principal – the organism feeds upon
a food supplied by a human, and the products of the organism's metabolism are used for human benefit. There are two categories of end-product.
The first category we call primary metabolites. These are the metabolic
products that the organism makes to provide for itself, to grow and to
develop: examples of these are vitamins and proteins. The secondary
metabolites are, if you like, the wastes that the cell discards. Many molecules that are important to us are waste products to an organism, so we can
culture the microbe and harvest what it makes. One organism is a small
rounded fungus cell, *Saccharomyces*. It feeds on sugars and its genes code
for enzymes that reduce starch to still more sugars on which it can feed. As
it feeds, it grows and, as it grows it releases carbon dioxide (just as we do)
and liberates waste into its surroundings. As the sugar is consumed, the
waste accumulates. *Saccharomyces* is a yeast, and its chief waste product is
ethanol – the alcohol of beers, wines and spirits. If the yeast is grown in a
spongy mass of starch, then the carbon dioxide will bubble up and make
the mass rise. When it has risen, cooking will drive off the alcohol and we
are left with bread. As bread needs bubbles of carbon dioxide to rise properly, there are other ways of achieving this end. Baking powders (such as
sodium bicarbonate) release carbon dioxide when heated and this allows
one to make an 'instant bread' without the need to wait while the yeast
works its magic. The bicarbonate is reduced to carbonate and – as sodium

carbonate is soda – the product is known as soda-bread, and it does have a slight taste of soda. Although soda-bread is a well-established bakery product, it is instantly recognizable as a very different product from bread made with yeast. In no respect is it a viable substitute. This reminds us of a crucial benefit of microbial fermentation – the subtle changes in flavour and texture that the traditional fermentation alone can achieve. The varieties of bread, all of them refined over centuries to optimize the end-result, are inimitable and most are testimony to the industry and inventiveness of our forebears.

Bread results from the growth of yeast in a semi-solid dough. If the yeast is grown in a liquid brew, then the carbon dioxide can escape to the air and the alcohol remains. The result is wine. If the fermentation is done in a

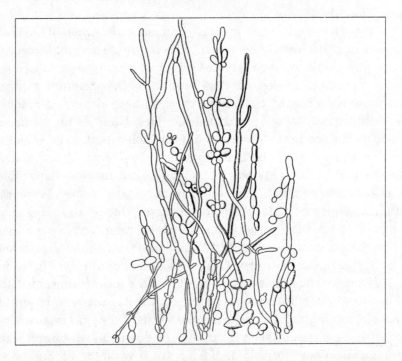

Harnessing yeast to work for humankind. Some yeasts grow as filaments, most grow by budding. In this form we can see both. This illustration was published in France in a nineteenth-century textbook on brewing. Relatives of this fungus occur naturally on the skin and on the surface of fruit. Some are harnessed to produce bread, beer and wine; other types can cause outbreaks of thrush in human patients, young and old.

closed container, then both the alcohol and the carbon dioxide will be trapped in the final product – and that is how we obtain champagne. There are other fermenting organisms that help make different kinds of beers, such as the lambic beers of Belgium. These beers are made by running the wort – the stewed mix ready for fermentation – into large, shallow copper baths. The microbes float in from the surroundings and the beer is produced through their natural fermentation. One brewery manager tells me that he was obliged to replace the ancient tiles on the roof of his brewery and install clean modern roofing instead. Next spring, when the beer was due to be made, they prepared their wort – and nothing happened. The spores of the beer-making microbes had been lurking in those ancient tiles, so they had to mount an expedition to find where they had been dumped, and reinstate them. Once the old roof was safely back in place, the fermentation worked as normal. If alcoholic beverages are left to ferment further in the presence of air, acid-forming bacteria get to work. They break down the alcohol to form acetic acid, and the result is vinegar. Malt vinegar is made from beer, but there are wine vinegars too, made by trickling wine through chambers where oxygen can dissolve in the liquid, so that the acetic acid bacteria can get to work. Depending on the environment in which the alcoholic liquid is treated, a whole range of different end-products can be made. One such is acetaldehyde, useful in the chemical industry.

The microbe cells are simply living, growing, respiring and producing wastes – but the way in which this proclivity is harnessed allows us to make a whole range of products from wines and beers to breads and cakes: all from a single, simple microbe. Many of our traditional foodstuffs involve the harnessing of microbe life. Tea, coffee and chocolate are all made with the intervention of microbes. The origins of chocolate lie in the Aztec lands. The people consumed large amounts of a drink they called *chocolatl*. Christopher Columbus had brought chocolate to Europe in the late 1400s, but nobody appreciated its significance. A chocolate drink soon became fashionable in the Spanish court. They made it by mixing the cocoa beans with cinnamon, nutmeg and sugar. The rarity of cocoa beans in Europe led to the secret of its preparation being closely guarded for a hundred years. The essence of chocolate is cocoa, extracted from the beans of the small tropical American tree *Theobroma*. The cocoa beans are collected and left to ferment for about a week; the sugars in the pulp turn to alcohol and the full flavour of chocolate develops as the fermentation proceeds.

The theobromine in the chocolate has a calming and stimulating effect on the consumer.

The coffee tree produces beans that ripen to a reddish colour and which look a little like cherries. The berries are opened by a pulping machine and the beans are fermented before they are sun dried. There is a cheaper way of processing, where the beans are simply dried whole – without the microbial fermentation. The result is 'hard coffee' and it is a of lower quality than the traditionally fermented coffee. Hard coffee is the type most usually produced in Brazil. Among the highest-quality coffee beans are those of Jamaica and Central America. The fermentation of the beans by microbes is the key to the traditional processing of the best coffee in the world, but many producers do not realize this. They omit it as it is seen to be a time-wasting procedure. Even tea benefits from a phase of fermentation. The tea leaves are crushed and spread out in layers, when they are left to ferment for a period of time which depends on the variety of tea. In producing conventional black tea leaves, the leaves are broken and cut to pieces, and are left to ferment for about two hours. Chinese teas are fermented for a shorter time – indeed green tea is not fermented at all, but is heat treated so it does not alter greatly from its composition when picked. Pouchong tea is fermented briefly – perhaps half an hour – and oolong tea is left to ferment for about an hour. The length of fermentation time is proportional to the strength of flavour: the more fermentation, the stronger the taste. Even coffee has a fermentation phase, in which the beans are left soaking for two or three days to free them by digestion of the pulp that surrounds them when fresh.

Milk is fermented in a host of ways. Cheese is now the most familiar, but many forms of fermented milk have been made for thousands of years. Some of them contain alcohol, because yeast is involved in the fermentation process. The Russians know *kefir*, a fermented milk containing alcohol and carbon dioxide bubbles, and *kumis*, alcoholic fermented mare's milk. In the Middle East we find *leben*, which has been made from the milk of bison and goats. Scandinavia has a huge range of yoghourts, including some stringy milks that contain heavy growths of streptococci – bacteria growing in long strings – which confer their property on the milk. Yoghourt itself comes in many varieties, and it is strange to reflect that (although it is very well known in the modern Western supermarket) yoghourt was hardly known outside the ethnic groups that produced it as recently as the 1960s. Sourness inevitably develops if fresh milk is left to stand in a warm place, because it contains small numbers of *Streptococcus lactis* bacteria which break down lactose in the milk to form lactic acid. The acid environment prevents most other bacteria from growing (and before long it even stops the *S. lactis* from continuing to reproduce). However, there are also lactobacilli in the milk, and these small rod-shaped bacteria grow and multiply rapidly. The acidity of fresh milk is about pH 6.8 (virtually neutral). By the time the *Streptococcus lactis* growth is at its height, the pH falls to around 4.0, and as it goes on falling these bacteria die out and the lactobacilli come to the fore until the acidity falls to around pH 3.0 (which is really quite strong, as acids go). As the fungi develop, the pH rises again towards 5.5. Careful control of the maturation conditions is needed if this sequence is to be properly maintained. The production of yoghourt requires a culture of specific streptococci; *Streptococcus bulgaricus* is the species most commonly used to make yoghourt, as you will see on the label. If cheese is being made, then the souring milk is raised to 30°C (86°F) and some rennin is added. Rennin is an enzyme found in rennet which sets the protein in the milk to form curd, a jelly-like mass. It is this that matures to become the final cheese. Soft cheeses are usually ready in a few weeks, although hard cheeses can take a year to mature. The blue cheeses are semi-hard varieties, which are spiked with metal rods carrying fungus spores. The fungi colonize the cheese, producing the attractive blue–green veining so loved by the connoisseur. Among the fungi that grow in cheese are *Penicillium*, the same genus that gave us the world's first antibiotic, penicillin, in the 1940s. Some cheeses become utterly rancid before consumption, and the reason we speak of people sometimes having 'cheesy feet' is because the same bacteria proliferate in semi-soft cheese as those that grow in the sock-enclosed environment of a damp foot.

In some instances, microbes themselves are even eaten as food. In the English-speaking world, salted yeast pastes are marketed as a by-product of the brewing industry. These are brown savoury spreads sold under such registered trade names as Marmite, Yeastrel and Vegemite, and adherents of each brand regard theirs as the product of choice. Although they are rich in proteins and vitamins of the B group, these are an acquired taste, and advertising agencies often say that 'you either love them, or you hate them'. In the Far East, the consumption of microbes reaches a more dignified plateau. Cultures of filamentous algae of the genus *Nostoc* are eaten in China as *to-fa-tsai* (literally, 'hair vegetable'), and in Japan there are unique colonies of bacteria which are found just beneath the surface of the soil on the volcanic slopes of Mount Asama. These bacterial masses are collected and eaten as a broth named *tengu*. This is also the name of the deity said to reside on the mountain, so the eating of this fabulously rare dish commemorates an ancient Japanese god. In the West we have begun to grow fungi in vats and texturize the result so that it has a similar feel to meat. This is increasingly popular with vegetarians, though one might think it better to make the texture as different from meat as possible. Quorn is a popular example of these new foods, and they offer a good range of proteins and vitamins while being free of fats and cholesterol.

Although we are developing new foods, we still surround ourselves with tradition whenever we can. That most ancient of products, leather, is also traditionally cured with the aid of single-celled microbes. The animal hides were soaked in pits of liquid and bacteria (such as *Bacillus erodiens*) could penetrate the hides and dissolve away the cellular components that the tanner wishes to eliminate. What the bacteria do not attack are the fibrous components of the hide and the result is that the hide is processed into supple leather by agencies too small for us to see. Much the same process has been used for centuries or more in making linen. The stems of the flax plant (currently grown to provide linseed oil) and hemp (better known as marijuana) were harvested and gathered into bundles which were steeped in soaking pits. The stems are a rich source of fibres – ideal for making textiles and ropes – but these are held together in the plant by pectins and other gluey compounds. The most effective and easy way of separating out the fibres was to use bacteria which removed the pectin and glue. In the soaking pits, a sequence of bacteria went to work: first the oxygen-breathing aerobic organisms, which quickly digested the softer tissues. As oxygen levels in the water fell, butyric acid bacteria started to come to prominence and they would go for the tougher glue-like compounds holding the fibres together. The result was that clean, glistening fibres could be recovered from the fermentation tanks, ready to be washed clean and woven into linen cloth, or spun into ropes. Even the ancient dye woad was produced by microbial fermentation of leaves. It is widely said that the woad plant, *Isatis tinctoria*, contains a blue dye. Not so. The dye, which is better known as indigo (and can now be made by chemical synthesis), is produced in a mash of the leaves when they are left to soak and then to ferment. It is microbes that produce the colour, and indigo was a prehistoric dyestuff of enormous importance.

Microbes have had a strategic importance in helping the antagonists in war time. During World War I, acetone was in short supply in Britain because of the German submarine blockade. A supply of acetone was vital for the production of cordite, the explosive used to propel shells from guns. The minister responsible, David Lloyd George, was told by a bacteriologist named Chaim Weizmann that a microbe could come to the rescue. Weizmann demonstrated that *Clostridium acetobutylicum* contained the genes that could initiate the production of acetone from crude sugar solutions, and this revelation was of crucial importance in maintaining the production of armaments. Without this microbe, the outcome would have been very different. The Germans, meanwhile, were running short of the glycerol that they needed to produce their explosives – and here too industrial fermentation was able to rescue their war effort from collapse. They found that the yeast *Saccharomyces*, if supplied with sodium bisulphite in their culture, could turn sugars into glycerol. Using this rapidly developed technique they were soon producing about 1,000 tonnes of glycerol per month. It is a little-realized fact that microbes helped fight the war, and without them the course of history could have been very different. There were further political repercussions, for after Lloyd George was appointed Prime Minister he found himself able to return the favour, and agreed with Weizmann over the establishment of a Jewish homeland in the form of the State of Israel. Rarely has a microbe been so politically active as *Clostridium acetobutylicum* was in World War I.

Even in the world of recreation we have found modern applications for microbes. The world-famous ski resorts of the Rocky Mountain rely on the length of time their snow lies, and research into modifying the weather (in order to lengthen the snowfall season) has never been successful. For many years they have tried to create artificial snow, but the sharp snow crystals produced by the machines were a poor substitute and were not liked by skiers. Bacteria have again come to the rescue: their surface coating encourages the crystallization of feathery snow crystals, each using a bacterial cell as the condensation nucleus on which the snowflake starts to form. Since the bacterium was brought in, the resort operators have been able to increase the duration of their holiday season considerably.

Microbes are even being used to mine minerals. Some algae have the ability to bind to valuable metals, including gold and silver. These microbes could allow us to re-open exhausted mines. Microbes can reclaim heavy metals such as cadmium, lead and even uranium from low-grade ores. Bacteria have been harnessed for many centuries to enable industry to get at low-grade iron ores. The method was to load low-grade ores into pits and allow water to trickle slowly through them. Bacteria of the genus *Thiobacillus* quickly become established, as they have an ancient lineage which gives them a preferred substrate of elements such as copper, iron and sulphur. They oxidize them, and capture the energy for their own life processes. The result is that a solution of iron (or copper) sulphate leaches out in the drainage from the pit, and from this chemical compound it is relatively easy to extract the pure metal. One of the greatest benefits of using a microbe is that it does not require an external source of energy to fuel its reactions – so a fuel line or an electricity supply to power the process is unnecessary. That is not to say that no energy is needed at the factory, because the equipment still needs to be operated. In some designs, energy liberated by the microbes (through a gaseous by-product, such as methane) can be harnessed to make the entire factory self-sufficient. Energy is also needed to cool the reaction, because the energy that microbes release can be considerable. However, the cells power the chemical reactions by themselves, and this can bring about a considerable saving of operating costs.

One of the best-known biotechnology industries is the production of antibiotics. These are fascinating compounds, because they are produced by microorganisms as waste products and are used by us in the conquest of bacterial and fungal diseases. In recent years they have been grossly overused, but that is not to say that antibiotics themselves were ever a bad idea. The lives of countless people are saved each day in a manner that would otherwise be impossible. I have been the first to criticize the pharmaceutical companies for encouraging the sale of antibiotics on a grand scale, though my recovery from bacterial meningitis (due entirely to the timely use of penicillin) does put those objections into a clearer perspective. The actions of penicillin were first documented by Alexander Fleming in 1928, although he did little with the discovery at the time. Four years later a hospital doctor was the first to use Fleming's cultures of *Penicillium notatum* to treat ophthalmia neonatorum, a leading cause of blindness in newborn children at the time. Fleming was not involved in the medical uses of the drug until it had been studied by Sir Howard Florey and his colleague Ernst Chain at Oxford. The first patient whom they treated was a policeman dying of septicaemia. Injections of penicillin produced a miraculous improvement – but then the supplies ran out. All the penicillin ever produced had been injected into the patient. The medical team managed

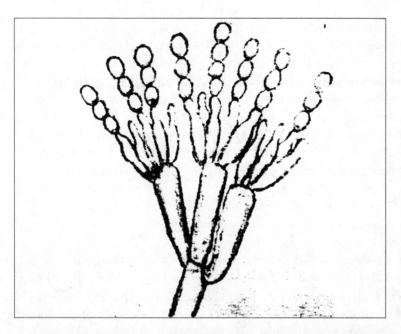

The penicillin mould which revolutionized medicine. Penicillin was the first antibiotic to be commercially produced. It has greatly reduced human suffering. This is the first picture of the penicillin fungus ever published. Twenty years before Fleming's discovery, *Penicillium notatum* was discovered by R. Westling in Sweden. This is a copy of Westling's original drawing. He found the fungus growing on a pile of rotting hyssop (a fragrant herb).

to recover unused penicillin from the patient's urine for a time, but in the end it proved impossible to hold the infection at bay and he died soon afterwards.

The large-scale production of penicillin began by culturing the fungus in large numbers of milk bottles, but the industry soon moved towards production in fermentation vats. Penicillin is now produced from a mutant strain of a different species, *Penicillium chrysogenum*. The wild form of this species secretes about 60 milligrams in a litre of culture. However, successive mutants have been studied. Two mutations increased the yield to 500 milligrams per litre, and another 18 mutations obtained in the laboratory increased that again to seven grams per litre – an increase in productivity of over a hundred times. A related genus, *Cephalosporium*, produces cephalosporin, and a group of filamentous bacteria – the actinomyces, found widely in soil – have given us streptomycin and tetracycline. These two groups of microbes have given us 4,000 of the 5,000 known antibiotics. In spite of this, only about 100 have ever been marketed, and two-thirds of those come from the single genus *Streptomyces*. Streptomycin was discovered in *Streptomyces rimosus* and chlortetracycline in *S. aureofaciens*.

Genetically modified microbes are used to make drugs such as interferon, insulin and human growth hormone. The insulin produced by microbes was not successful at first, because it was subtly different in structure to the human version and was unreliable in its effects. However, human growth hormone extracted from tissues has proved to increase the risk of other diseases (including the spread of Creutzfeldt–Jakob disease or CJD) so the purer form made by microbes may prove to be a life-saver.

Sometimes, we have taken advantage of a microbial product when it becomes painfully evident in the wrong context. Dextran is a mucilaginous substance and it is often formed wherever microbes are feeding on glucose. Sugar refineries (which extract sugar) and ham-curing establishments (which use it) are often bedevilled by slimy formations of dextran. It can make drinks undrinkable and food uneatable. If it spills on the floor it can make it dangerously slippery; if it forms in pipelines it can repeatedly block them. This polysaccharide substance is persistently glutinous – which can actually be an advantage, if used in the right place. It has been found that dextran is well tolerated by the body. It doesn't show signs of incompatibility with the body by triggering an allergic reaction or a rise in temperature. Its glutinous nature makes it the ideal substitute for a plasma transfusion, and dextran has become widely used to help boost the volume of blood in accident and emergency patients. It doesn't suffer from being heat sterilized, and it can stay in storage at room temperature for prolonged periods without special facilities. So we use it as a blood expander. A solution of dextran, carefully balanced to suit the body, can be transfused into people whose blood volume has been reduced through accident or surgery. The bacterium used for the production of dextran is *Leuconostoc mesenteroides*, whose genes code for the production of the most suitable molecular form for use in medical emergencies.

One class of microbial products has become well known, and this is the enzymes. Enzymes are responsible for initiating the chemical reactions in all living things, and are medically and commercially important. Commercial television wouldn't be the same without enzymes. Washing powder advertisements assume that anybody caring for the household's clothes shouldn't be trusted to wash anything in water that's hotter in degrees centigrade than the measure of their own IQ. To the public, the meaning of enzymes remains a mystery. Enzymes are complex proteins the production of which is coded for by genes within the cell; they trigger chemical reactions. If you burn glucose in a fire, then the energy of combustion is released in a burst of flame, and that's it. Consume it as an energy source in a living cell and the reaction takes place as a controlled sequence with enzymes regulating each stage of the process. Living cells use enzymes to digest their foodstuffs. The conventional way of naming enzymes is to set down the scientific name of the substance on which they act, adding -ase at the end. The enzyme that helps break down the sugar called maltose, for example, is maltase. Protease would be an enzyme

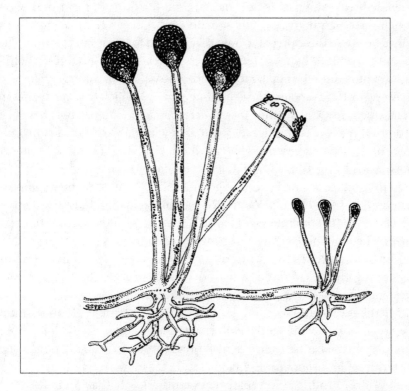

A common bread mould used to produce enzymes. The mould which forms on stale bread is often *Rhizopus*. The black spores form inside round sporangia on the ends of fine thread-like hyphae. The sporangia suddenly collapse when ripe, and this catapults the spores into the air. *Rhizopus* is commercially grown in factories where it produces maltase, an important enzyme used to produce sugars from starch.

that breaks down proteins, and it is these that you find added to washing powders. In this application, they help to break down what are delicately called 'body stains'. In industry they are used in the recovery of silver from used X-ray films, where they dissolve away the gelatine of the film emulsion.

You will probably be acquainted with the moulds *Aspergillus* and *Rhizopus*, because they commonly occur on stale bread. Both are used to produce amylase and maltase, respectively. Amylase breaks down starches to sugars, and you will find it widely used in the brewing and baking trades. Some commercial brands of dried baking yeast include amylase in the mix, as well as resting spores of the yeast itself. When used to make bread, the amylase breaks down some of the starch in the flour to form sugars, on which the emerging yeast cells start to feed. This gives their growth an added boost, so these preparations are marketed for their fast reaction. The maltase produced by *Rhizopus* changes the relatively unusable carbohydrate maltose into the much more useful glucose. Strains of *Aspergillus* also produce proteases and pectin-degrading enzymes which have many uses in the textile trade. Cobalamin (the modern name for vitamin B12) has been produced from yeast for decades, but is now usually extracted from cultures of *Propionibacterium*, whereas riboflavin (vitamin B2) is commercially made by the fungus *Ashbya gossypii*. There are many other microbes used to produce amino acids that go into food. Many of them are important dietary supplements. We need 20 amino acids for a healthy diet, of which we – or the microbes within our intestines – can produce 12 for ourselves. The remaining eight must be supplied in the diet, and we now have cultures of microbes that can produce them all. A great range of acids (including citric acid, used in fruit drinks) is produced by microbes at greater efficiency than a purely industrial process.

Microbes can produce food flavours and even perfumes. Monosodium glutamate is widely thought of as a synthetic food enhancer, but it has been produced in the Far East since ancient times through a fermentation by the bacterium, *Corynebacterium glutamicum*. In the modern era, some 350,000 tonnes are produced annually by cultures of that organism or *Brevibacterium flavum*, of which there are some highly productive strains. In the perfume business, one of the most highly prized ingredients was always ambergris. It is produced from the fermented excrement of the sperm whale, and the lumps of ambergris found floating in the sea are now reckoned to be made of whale faeces compacted around a central mass of the indigestible remains of squids (including their beaks). Fresh ambergris is almost black in colour and smells most objectionable, but as it floats in the sea, and the microbes slowly ferment its components, it fades in colour and develops a most appealing aroma. Whether its origins are known to those who have used it over the centuries is not clear, though I wonder whether it would have been greeted with such rapture if they were.

5

Cells against pollution

IN MY TEENS I went camping with a friend by a waterfall in the mountains. Alongside was a chuckling stream which burbled its way down across the gravelly bed-rock and spattered against the slanting stones in the sunlight. We crouched down and drank from cupped hands. The water was cool, clear and reviving. We splashed it on our faces, rubbed it through our hair. And then we trudged up the wooded slopes, the shadows of dappled leaves dancing around our feet, to see where the stream came from. Around a bend in the stream came the answer: propped in the water like a garish sacrifice was the half-rotten corpse of a sheep, hollow eye sockets staring to the sky, flies buzzing around the cavity where its brain had been, with strands of putrid intestines and tufted strands of rotting wool waving in the water. The once-crisp taste of the water in one's mouth seemed to acquire a different flavour after that.

Did we become ill with dysentery? Was the water poisoned and full of germs? Of course not. The water we drank, just a short distance downstream, was perfectly pure. We owe the purity of water not to the fact that it has never been touched by microbes, but because it is populated by them. It is the presence of microbes in water that renders it free from germs. Microbes are such a varied bunch, but the overwhelming majority are harmless to human life and are specialized for organizing our environment and keeping it running. There is no 'balance' of nature; the equilibrium that we see results from continuing warfare between disparate groups. Let us put a lens on that mountain stream, to witness what happens. The environment depends on the microbes of decay to remove dead matter from

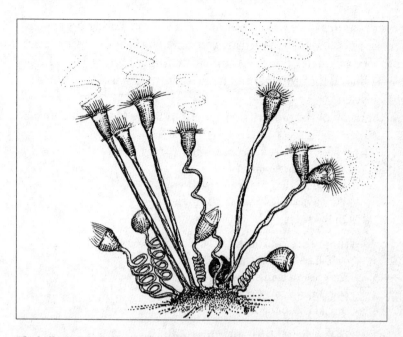

The bell animalcule _Vorticella_ purifies water. The presence of microbes in water may make you conclude that the water is bound to be dangerous. On the contrary, it is microbes that purify water. These tiny organisms waft a current of water through the cell with the aid of their whip-like cilia. Bacteria in the water are trapped and eaten as food. There are trillions of these organisms at work inside a biological water filter, whether at a large water works or even, on a smaller scale, in your aquarium at home.

the remains of that sheep. They proliferate in vast numbers, using enzymes to break down the tissues, and they continue to reproduce, building their own cells from the raw materials that they recycle. Immediately downstream from the sheep, the water is rich with these bacteria as the flow washes them away. Attached to the rocks and pebbles are other microbes, the next stage in the recycling on which nature depends. They feed by whisking a continuous current of water through the opening of the cell, the flow maintained by the energetic strokes of thread-like cilia which beat in concert and look under the microscope like waves of wind across a field of wheat. These water-cleansing microbes feed on the very kind of bacteria now multiplying within the sheep. The greater the supply of bacteria, the more rapidly these ciliated microbes reproduce. As a result of their activities, the water a few metres or yards further downstream has been cleansed of the bacteria and is once more clear and wholesome. Larger organisms feed on the ciliates, and they in turn are consumed by fish; so – by the time some enthusiastic angler downstream is catching a plump fish – he is holding in his hands the protein from the sheep upstream which nature has efficiently recycled.

You might think of the old British rhyme 'On Ilkley Moor', which runs like this:

> Wheear 'as tha bin sin ah saw thee,
> On Ilkla Moor baht 'at
> Wheear 'as tha bin sin ah saw thee?
> Wheear 'as tha bin sin ah saw thee?
> On Ilkla Moor baht 'at (repeat chorus)
>
> Tha's been a courtin' Mary Jane
> On Ilkla Moor baht 'at
> Tha's been a courtin' Mary Jane
> Tha's been a courtin' Mary Jane
> On Ilkla Moor baht 'at
>
> Tha's bahn t'catch thy death o'cowd
> On Ilkla Moor baht 'at
> Tha's bahn t'catch thy death o'cowd
> Tha's bahn t'catch thy death o'cowd
> On Ilkla Moor baht 'at
>
> Then we shall ha' to bury thee
> On Ilkla Moor baht 'at
> Then we shall ha' to bury thee
> Then we shall ha' to bury thee
> On Ilkla Moor baht 'at
>
> Then t'worms'll cum and eat thee oop
> On Ilkla Moor baht 'at
> Then t'worms'll cum and eat thee oop
> Then t'worms'll cum and eat thee oop
> On Ilkla Moor baht 'at
>
> Then ducks'll cum and eat oop t'worms
> On Ilkla Moor baht 'at
> Then ducks'll cum and eat oop t'worms
> Then ducks'll cum and eat oop t'worms
> On Ilkla Moor baht 'at

Then we shall go an' ate oop ducks
On Ilkla Moor baht 'at
Then we shall go an' ate oop ducks
Then we shall go an' ate oop ducks
On Ilkla Moor baht 'at

Then we shall all 'ave eaten thee
On Ilkla Moor baht 'at
Then we shall all 'ave eaten thee
Then we shall all 'ave eaten thee
On Ilkla Moor baht 'at

That's wheer we get us oahn back
On Ilkla Moor baht 'at
That's wheer we get us oahn back
That's wheer we get us oahn back
On Ilkla Moor baht 'at

Thus, the worms eat the deceased, ducks consume the worms, and the villagers in turn eat the ducks. They eat the deceased and get their own back. It is a perfect understanding of recycling in nature. Now we can add to this vivid reminder of nature's capacity the crucial part played by microbes. They are what power the churning energy of life on our planet, just as they cleansed the water that ran through the rotting corpse of a sheep in the mountains.

A familiar example often in the news is the pollution of the Mediterranean Sea. The Mediterranean is surrounded by the oldest developed communities in the world. Billions of gallons of industrial pollution and human waste pour into it every day. It has been dubbed the most polluted sea in the world. Yet look at the famous holiday beaches and you find a different story. The sea itself is renowned for its sparkling brilliance. The underwater visibility is legendary. Holiday photographs show the tourists enjoying the clear blue waters. The Mediterranean shows how nature's microbes work against human pollution. Incalculable numbers of microbes are incessantly at work, consuming the wastes and recycling them. In pollution studies, microbes are too often omitted from the equation. As long as the Mediterranean microbes are left to their own devices, they will naturally decontaminate much of what we keep pouring into it. As microbes still have metabolic systems that echo the conditions of the primitive earth, many of them rely on chemicals that we think of as poisons. They may relish a diet of cyanide, for example, or subsist on ammonia. It is true that we pour huge amounts of human excreta into the limpid waters, but the Mediterranean has been dealing with faeces since long before people started to show an interest in the problem. The run-off from the land has always contained excretory residues from animals on dry land, and the huge populations of sea creatures churn out incalculable amounts of excrement every second of the day. There are microbes in abundance which feed on these and recycle them into a form that can be utilized by larger forms of life. If you could rid the sea of its microbes, it would soon become a polluted and poisonous chemical soup. As a result of the microbial community, the wastes from the Bay of Naples end up on plates in the form of sea food. All life in the Mediterranean depends on the continued input of chemicals from the land. If we load estuaries with high levels of mercury, for example, we will continue to run risks of poisoning people. If we allow the waters near atomic power stations to become cont-aminated with nuclear wastes, there is going to be a hazard to children playing on the beach.

Already we have inflicted these threats on the public. The World Health Organization has been studying poisoning in remote areas of Brazil, and they conclude that people can suffer nerve damage even if the levels of mercury are below those widely regarded as safe. The tribal peoples in the Amazon who live downstream of gold mines have been shown to be accumulating mercury in their bodies. The researchers tested their hair and found traces of mercury throughout the people. The immediate conclusion was that the poisonous metal was being released into the environment by the prospectors panning for gold, and this was substanti-ated by the discovery that 130 tonnes of mercury are released into the environment by the mining operation every year. The mercury was used by the estimated one million prospectors hidden in the forests, and hoping to make their fortunes. Case closed? Not quite – just as the Reverend John Needham discovered, you need further investigations to confirm any such preliminary finding.

The main problem was that mercury poisoning was being found hun-dreds of miles downstream, so measurements were made of the levels of mercury in the water. It was soon found that the amount of mercury from river water samples taken near the miners was the same as that taken 300 kilometres (nearly 200 miles) away. The total amount of mercury in the environment was far higher than the total amount of mercury used by the gold prospectors. It was not coming from the mining operation at all, but from the stripping of forest trees. The cutting down of the rain forest was allowing soil to run off in rain storms, carrying a muddy slurry into the rivers and washing downstream. The mercury was naturally present in the soil. It had been tolerated by the microbes well enough, and was held in the soil out of harm's way, but soil erosion washed it all into the water and poisoned thousands of people. The WHO scientists measured the levels of mercury in villagers' hair and then gave them visual and dexterity tests. For many years a safety level of 50 parts of mercury per million (p.p.m.) had been accepted as 'safe', but these new investigations revealed that villagers exposed to lower levels of mercury had difficulty in seeing fine details on eye-test charts. The problems were proportional to the levels of mercury to which the individual had been exposed. In the short term, the researchers have advised the villagers to stop eating the predatory fish which accumulate high levels of mercury from the many stages in the food chain that went before. Of more long-term importance is a halt to the destruction of the rain forest. Brazil is destroying between 25,000 and 50,000 square kilometres (about 20,000 square miles) of rain forest per year. However, it is clear that the microbes in the soil of the forests have found a way to dispose of mercury which our disruption of the ecological system can disturb. Meanwhile, the easy answer – that mercury used by greedy gold prospectors must have been poisoning the people – is not tenable. It is the cutting down of a stable forest ecosystem that has liberated naturally occurring mercury into the rivers. We owe our environmental safety to the systems of nature, and we disrupt them at our peril.

The WHO obtained the figure of 50 p.p.m. from studies of the greatest of all the tragedies caused by mercury in the environment. This was the terrible outbreak of Minimata disease in Japan. From the 1930s, the rapidly expanding chemical industries at Minimata, Japan, used the nearby bay as a dump for all their wastes. High levels of mercury were poured, unchecked, into the bay. The capacity of the confined ecological systems to dispose of the mercury was soon overloaded, and mercury began to appear in the fish. The larger the fish, the higher the concentration of mercury. During the 1950s, a new epidemic was reported: the fishing communities clustered along the shore were developing a range of neurological symptoms. They developed weakness of the limbs which caused some of them difficulties in walking, and tingling of the extremities. Some became deaf, others began to lose their sight. Many developed indistinct speech, and a few started to behave strangely, sometimes bursting into laughter without apparent cause. During the 1960s, the cause was identified as mercury poisoning of the environment, and the offending industrial plants were closed down. This tragic experience set accepted limits to future mercury contamination, but the recent work in the Amazon shows that these are too lax. Clearly, more research is needed on the way microbes tackle poisons such as mercury in the environment – and we should take greater care in future before we disrupt stable ecosystems. The reasons against the destruction of the rain forests are already well rehearsed, but the release of toxic stores of heavy metals into the environment may be one of the strongest new arguments of all. Mercury can be found in many areas of the world. For instance, levels of methylmercury (responsible for the Minimata disaster) are already causing concern in lakes in North America and Scandinavia.

All around the world, microbes are always at work, their genes fitting them to recycle organic wastes. One of the most popular methods of disposing of the huge amounts of garbage we discard each day is landfill. It seems to make perfect sense: there are great gaping chasms cut into the surface of the earth where we have quarried rocks and extracted ores, and low-lying areas subjected to flooding which might benefit from being raised over the course of a few decades. Filling these depressed areas with loads of unwanted wastes seems the ideal way to answer two problems at once. But what happens to the waste when it is buried? It seems to have simply vanished from sight – but it is just starting out on the most active and productive phase of its existence. From the moment it is discarded, any scrap of moist organic matter in the refuse is colonized by microorganisms. Not just food waste, but paper, wood, packaging and cartons are all reduced to a soft pulpy mass. Most metallic wastes are broken down and reduced to an oxidized ore. As plastics are synthetic, there are no microbes that can yet use them as a food source, but many of the plasticizers that make them flexible tend to break down. Much of the plastic waste becomes friable and becomes incorporated in the general organic mass inhabited by the microbes. What happens next? Much of our environment has been made by uniting such things as water and carbon, and we have become accustomed in this book to seeing organic matter broken down mostly to form carbon dioxide and water. That is what happens in your bonfire; that is what will happen to you. However, it does not happen in the landfill. There is too much matter to decompose, and too little oxygen to do it properly. This has little effect on the microbe communities, which seize the opportunity, so that the types that multiply rapidly are those that do not need oxygen. These are the anaerobic bacteria, some of which are related to types that were around before there was any oxygen in the air. Instead of carbon dioxide, these produce a different gas – methane. The great beds of natural gas that modern society is draining are left over from the breakdown of organic matter in an oxygen-free environment by microbes adapted to the task, and this is what happens (on a much-reduced scale) in the modern landfill.

Some local authorities harness the methane and use it as a fuel. In Eastern Germany there are old landfill sites. They have been covered with soil and planted with coniferous woodland, making a pleasant environment for local communities which betray nothing of their history – apart from the occasional pipe headers that project from the ground in the occasional forest clearing. These are joined to a network of perforated pipes spread across the refuse before the dump was closed over with earth. As the garbage breaks down and the microbes produce their methane, the gas is collected in the tubes and used as a fuel. Nations that do not do this are posing several problems. First, there is a large amount of waste gas which could be usefully consumed. Second, escapes of the gas run the risk of causing fires. (Methane is highly inflammable, and sometimes bursts into flame without outside sources of ignition. Bubbles of methane sometimes escape from soggy marshland or peat in pools, and when they catch fire spontaneously they produce a flickering ghostly flame visible over considerable distances. This is known as 'Will o' the Wisp' and is well known to people living in swampy regions; it is one of the origins of legends about ghosts, and may start to be reported over land-fill sites once more.) Third, the methane is a powerful greenhouse gas and adds considerably to the levels of air pollution with which our descendants will have to contend. We should recognize the value of domestic and industrial wastes. Most can be degraded by bacteria and harnessed as an energy source. If we learn to co-operate with the microbial world, we will create a new source of renewable energy, and perhaps find other ways of re-using materials that are currently discarded. A common example is the composting of waste. Microbes can break down organic waste to form compost, and this is the ideal soil conditioner. In this sense they are only doing what comes naturally. Much of our civilization is based on controlling microbes, and stopping them doing what they might otherwise do. When it comes to waste disposal, we would do better to co-operate with this invisible world.

The savings – in reclaimed materials, in reduced transportation costs, in lower pollution levels and in the creation of new products – are potentially very great.

We have long employed microbes – without realizing it – to purify our drinking water. The ciliates that purified the mountain stream grow naturally on the stream bed, and it is these that have been used to treat drinking water; indeed, their existence is still known to surprisingly few people. One of the traditional methods of treating water is the slow sand filter, in which the water supply is run onto the surface of a vast shallow pool containing a layer of sand on a bed of gravel and perforated tiles. Once the system has stabilized, the water that runs out from beneath the sand is freed from the germs that it originally contained, and a handful from the sand layer feels slimy to the touch. It was believed that this was the organic matter in the water which had been trapped by the grains of sand.

That isn't possible. The chief objection to that idea is that the gaps between the sand grains are far larger than the germs that are supposedly being filtered out. A grain of sand is a million times bulkier than a bacterium; in no way could you hope that a layer of sand could filter out the germs. You could make a case that they simply settled out onto the surface of the sand, but the slow sand filter only works when the water is flowing down through the bed, and if it is flowing it is not likely to allow bacteria to settle out. It is the microscope that provides the answer. Each grain of sand in a filter bed becomes rapidly colonized by the ciliated organisms that feed on bacteria. As the water moves down through the layer, these beautiful little iridescent cells are drawing a vortex of water through the mouth of their cells, removing each little bacterium and digesting it for food. As the ciliates multiply, the bacteria are systematically eliminated from the water. *Vorticella* is typical of the organisms that purify contaminated water by grazing on bacteria. It has a bell-shaped body on a spiral stalk. This stalk can suddenly contract, and then it slowly unravels itself in a different orientation. It does that first when it has exhausted all the food in that particular compartment of water – it jerks in, and then comes out in a different direction – and the response is also used to escape from danger. If some blundering water flea comes by, the stalk also jerks back in, and the water flea, with any luck, is denied its meal. *Vorticella* has a ring of cilia around the mouth of the bell, and these circulate a current of water into the mouth of the cell and out again. Any unwanted particles (small sand grains or dust particles) are ignored by the cell; but anything alive, such as a bacterium, is taken in by *Vorticella* and consumed. It passes down the gullet and into the cytoplasm. Any residues are ejected when digestion is complete.

If the water that is filtered is too rich in bacteria, then in time the sand layer needs to be replaced: it can literally become clogged with a growth of *Vorticella* and the other organisms that accompany it. Long experience has taught plant operators exactly which sort of replacement sand to use, how to adjust the water supply, and how deep the sand layer needs to be. Shortly after the British water industry was privatized, one sand filter bed due for refurbishment had a tractor drag out one corner of the sand layer rather too thin for optimum efficiency. The result was that a large volume of water was left contaminated. An experienced craftsman would have been able to avoid that.

Oil spills are among the most visible and damaging examples of pollution in the modern world. The television images of oil-bound bird life and struggling marine mammals travel around the world and strong feelings of revulsion arise. The tenacity of oil is well known to all of us, and its eradicable nature is familiar to everyone who has tried to shift oil from the hands after fixing an engine, or from shorts after sitting on a contaminated beach. But then, microbes have been dealing with oil for hundreds of millions of years longer than we have and, as you might anticipate, they cope with it better, too. As oil is an ancient substance, there are many types of organisms that can metabolize it. Among the bacteria are species of *Actinomyces* (the genus that gives us antibiotics), *Mycobacterium* (some other species of which cause tuberculosis) and *Pseudomonas* (some organisms in this genus occasionally cause hospital infections). As you might expect, some of the ancient sulphur bacteria are known to degrade oil, including *Desulphovibrio,* and a list of others. At the time of the Gulf War against Iraq, it was widely said that the leakage of oil into the seas would poison the Gulf for centuries. I did not subscribe to that view, and for very clear reasons. Oil has bubbled to the surface in the region for millions of years, and this is also an area where the seas are warm (which encourages bacterial growth) and shallow (which helps provide a better supply of oxygen than is possible in deeper waters). There are huge communities of organisms already in the area, where they have used oil as an energy source since long before the existence of humans. The natural leakage of oil in the Gulf was known to ancient peoples. It has been burned as a fuel and, where the thinner oils evaporate leaving behind hot deposits of asphalt, the local communities have long had a tradition of making models before the asphalt cooled into solid rock. To this day, visitors are sold black ornaments which were shaped by hand while the asphalt was still pliable and soft.

The sheer extent of the flooding of the Gulf with oil beggars belief. The spillage in Alaska from the Exxon Valdez in 1988 poured 38,000 tonnes (11 million gallons) into the sea. It contaminated over 2,400 km (1,500 miles) of coastline and the company were faced with a clean-up bill of $3 billion, together with a fine amounting to $5 billion. In Europe, the Torrey Canyon disaster off the Cornish coast in 1967 released a total of 121,000 tonnes (35 million gallons), and the Amoco Cadiz wreck on the French coast in 1978 had topped that with 225,000 tonnes (65 million gallons) emptied into the English Channel. The full extent of the Gulf disaster will never be known, but over 345,000 tonnes (100 million gallons) flooded the sea. Great flocks of cormorants and curlews, oystercatchers and redshank were seen floundering in the water, while colonies of manatees and groups of turtles were poisoned, and porpoises began to die. Yet by the time the reports were spreading around the world, the microbes in the sea were already beginning their work of recycling the oil back into harmless products. They brought about an extensive recovery of the oil-damaged Gulf in record time. To this day there are hardened deposits along the high-tide mark, where they join other deposits dating back thousands of years. But the oily deposits in the sea have disappeared, and the rate of recovery has exceeded expectations.

Even though this was a huge assault on the environment, it was not the greatest of all time. That distinction is held by a massive blow-out of the Ixtoc I oil well in the Campeche field in June 1989, which poured more than 500,000 tonnes (145 million gallons) of crude oil into the Gulf of Mexico. The resulting oil-slick extended to more than 650 km (400 miles). Needless to say, the microbe world came to our rescue to conquer the threat to the environment. In recent years we have seen further spillages on the west Wales coast, and the wreck of a tanker on the rocky coasts of Shetland. There are continuing spills in the oil fields from Mexico and right across the Middle East. Yet the most remarkable impression gained by visiting the sites of great oil spills is not the legacy of death and destruction, but an overwhelming sense of normality. We come along and spray the sea with detergents (which is now known to kill wildlife for miles around) and then cleanse sea birds in a massive effort of devoted humanity (although it is now clear that few of the birds ever survive long when returned to the wild). Yet microbes can remove the oil and restore the environment even if we do absolutely nothing. Most of them are types that are still poorly understood, and it is becoming clear that the best way we can help matters along is to spray compounds that help to nourish the microbes, rather than trying chemically to disrupt the system. This is a new approach, but one that allows us to interact with the forces of nature, rather than trying to subvert them. The Shetlands spill was the most remarkable example of rapid recovery. At the time of the disaster, official reports said that the areas would be devastated and that the ecological balance was lost forever. But the microbes are already at work. Within a week of the spill, much of the oil had been evaporated, and the rest was distributed through the ocean where it was digested by the microbes that have lived since time immemorial in the sea.

There are already some new techniques by which these microbes could be harnessed. One of them involves sinking floating oil spills to the sea bed by dosing them with silicone-treated fly ash. The ash resists the water, so it stays afloat until it binds with oil. Then, because the combination of ash and oil is denser than sea water, the mix sinks to the bottom of the sea. The incorporation of a mix of microbes in the ash means that the degradation of the oil starts quickly. This is an attractive idea at first sight, but sinking masses of oil poses problems for other users of the sea. It does mean, for example, that fishermen would be less than pleased to be dredging up loads of half-digested crude oil when they had rather hoped for fish.

One species of bacteria has been used to help clean out oil tankers. *Pseudomonas putida* has a natural appetite for oil, and some strains have been developed that can metabolize the most unappetizing compounds, including naphthalene and camphor. Cultures of this organism can be sprayed into empty tanker holds and it does indeed break down many of the residues. There have even been plans to use microbes to thin oil that is difficult to extract. Just as *Thiobacillus* can recover metals from low-grade ores, it should be possible to use cultures of bacteria to attack unusable tarry shales so that a thinner oil could be recovered. In some applications, water is run into oil fields as the oil is extracted, and it helps if this water is somewhat viscous. A microbe named *Xanthomonas* produces a mucilage known as xanthan, which confers exactly the right properties on water injected into oil fields, so to that extent the petroleum industry is using microbes already. But just how much is the microbe world using the petroleum industry? As you'd expect, microbes have lost no time in exploiting the way we use petroleum products. For decades it has been known that cutting fluid used on lathes in factories can clog jets, a surprising occurrence when the fluid is a carefully composed emulsion of clean oil in filtered water. Needless to say, the existence of an emulsion of water and oil is an ideal growth substrate for oil-degrading microbes, and it was their proliferation that caused the jets to clog. Special additives are used in aviation fuel, because microbes can grow in the fuel tanks of aircraft with potentially disastrous results. The problems caused by microbes in fuel tanks has been extensively investigated by military organizations, where the consequences can have strategic implications, and the oil prospector is aware of the way that microbes can eat their way through solid metal. A standard steel tube can be destroyed by microbial action within a year or two. It is important to consider this other side of the coin. All the time that we are seeking to utilize microbes to assist our aims, you can assume that the microbe fraternity are well placed to exploit our technology to serve their purposes. Microbes are great opportunists. They must be, or they would never have gathered together to form the great co-ordinated communities that we call 'trees' and 'people'. In the modern world, microbes are often able to adapt our innovations for their own benefit faster than we can harness them to serve our purposes.

The use of fertilizers (see page 149) and modern farming methods has given us prodigious success in food production, although intensive farming is now damaging the environment. Rather then returning composted organic material to the soil, we are dosing it with high-nitrate fertilizer. In consequence, the soil is oxidizing away and becoming thinner each year. In the fenlands around Cambridge, where I live, the peaty soil used to measure two or three metres (about ten feet) in depth. Most of it has oxidized away, and in some parts there are a few centimetres (an inch or two) of soil left, and clay is regularly ploughed up as farmers try to prepare the soil for a new growing season. There are vivid drawings of lush vegetation in caves in the Sahara, a reminder of an earlier era when that region was cultivated for food. These should be a warning of the way we are moving with the use of techniques of cultivation that pay no need to the importance of retaining soil in good organic condition. Microbes are a key component of the soil, and too little effort has been devoted to understanding how our long-term planning should take all such factors into account. The world now produces enough food to feed everyone, and it is not a shortage of food that causes starvation across areas of the developing world, but a shortage of goodwill. In Britain we have tried to become a producer of food, rather than an importer, and in the 1980s we finally reached that desirable end. This ought to have been the cause of rejoicing and celebration but, of course, the converse happened. We were castigated by Brussels for over-production, and the policy of 'set-aside' was introduced instead. Ever since that time, farmers have been paid large sums not to produce food. Efforts to move surplus supplies to areas where malnutrition is widespread have been much less evident.

Pests are among the greatest threats to the survival of farming and food production. Agricultural pests evolved with constraints imposed by the need to find new host plants to colonize. Modern agriculture makes the task easier: we plant row upon row of host plant, often extending right to the horizon. A pest species which might have expected to travel a kilometre (half a mile) or more before finding a new host plant will now find one, ready and waiting, right alongside. This is the greatest cause of outbreaks on farms across the world. Chemical sprays were always seen as the best defence against an invasion of pests, but more recent knowledge casts doubt on this approach. In the 1950s there was a belief that the USA could simply rid herself of insects, making the environment more pleasant for humans to occupy. It was an arrogant idea, and if it had proved possible it could have wiped out the greatest nation in the world. Insects are a vital player in the fight for life. They are a fundamental agent for pollination, which is vital for key plants to survive at all. They act as regulators of pests, by parasitizing many species that could otherwise prove intensely harmful to human society, and they are an important means for transmission of genetic information across considerable distances. Their larvae, like mag-gots in meat, are a first step in degrading an animal carcass, even if the dead remains are somewhere out of contact with the soil (and therefore not in immediate contact with many of the organisms of decay). A world without insect life would be hard to sustain, and human life would quickly be threatened were they all to vanish. There are doubtless many functions that insects carry out that are, as yet, unsuspected by science.

The first indication of the dangers of this approach came through the profligate use of DDT. This was not because of medical effects on humans, but because of problems for wildlife. As we progress along a food chain, we can model how easy it is for undesirable compounds to accumulate in the tissues. Let us envisage a pelican, for example, which consumes hundreds of fish; each fish will have consumed thousands of creatures in the plankton; each organism in the plankton may have consumed millions of microbes. If there is something undesirable in the microbes, then the increase in concentration as we pass up to the pelican would be in the order of one hundred thousand million. That is a simple idea of how a poisonous compound – if not degraded by microbes – could accumulate in the tissues at each and every stage. We are not so much faced with a food 'chain' exactly (all links in a chain are the same size) but something I prefer to call a 'food funnel', where a huge catchment is concentrated down to feed a new individual. Scientific data demonstrate how dangerous this can be. Published figures reveal that a concentration of DDT in water no greater than of five parts per ten billion can be funnelled into 23 parts per million in a bird of prey at the end of the line. That is an increase of around 500,000 times – 50,000,000 per cent, as a business would say. Even at this level, there were few effects on the medical state of the adult birds, but there was crucial interference with the way they precipitated calcium salts in creating egg shells. The shells were weak and easily broken. Few viable fledglings ever left the nest, and the survival of many species of common birds was threatened. Since DDT was banned, populations have increased once more. We should realize how close we came to losing some of the most important bird species in our environment. Malathion, which has been widely used in California to combat the Mediterranean fruit fly, is now known to have run off into rivers and damaged native fish stocks. Clearly, there are lessons that agriculturists still have to learn.

There is increased concentration of a poison through the funnelling seen in nature, but we also need to realize how toxic these compounds can be to humans. Kepone, a raw material used to make insecticides, was the cause of illness in a number of production workers at the factory in Hopewell, Virginia. Further investigation showed that it had contaminated the James River, which ran past the factory. Fish in the river were found to contain high levels of the compounds, and it was closed to fishing indefinitely. Clearly we need to be proactive in regulating the production of such compounds, and we must know more about the biodegradability of the new chemicals that we put into the environment. PCBs (polychlorinated biphenyls) are well known as an example of a widely used compound in the past which caused unexpected damage. The first major outbreak of disease was reported in 1968 in Japan, where over 1,000 people were poisoned by cooking oil poisoned by PCBs. It took 10 years before they were banned in the USA. Since 1978, there has been no further use of these compounds anywhere in the USA, but one-third of the American people still have more than 1 part per million of PCBs in their bodies.

Biodegradability is the key. DDT, surprisingly, has been found to be biodegradable in the laboratory. Like many such substances, it has enough in common with familiar organic compounds for the genes of bacteria to code for enzymes that are able to break it down. However, it is not an energy source, so organisms have not acquired a preference for it. In laboratory culture, bacteria provided with abundant sources of energy will break down the DDT molecule as they feed. In nature, where there is no metabolic benefit to be obtained by degrading the molecule, DDT is usually ignored by microbes. This is why it persists for so long in the environment. It is the chemical structure of a molecule that holds the key. Studies have shown a key difference between two herbicides, 2,4-D and 2,4,5-T. The former is biodegradable and vanishes from soil within weeks; the latter does not disappear, and can be found in contaminated soils many years later. The reason lies in the chemistry. The difference is a single atom in the formula. The composition of 2,4-D (2,4-dichlorophenoxyacetic acid) differs from 2,4,5-T (2,4,5-trichlorophenoxyacetic acid) by the addition of a single chlorine atom to the core of the molecule. This extra atom blocks the action of the enzymes that bacteria produce to break down the molecule, and so it persists in the environment.

Now that we can tailor molecules, we need to understand exactly what will happen to a compound once it has been sprayed around. The studies must take into account what happens downstream of the application. For instance, dieldrin (the insecticide) does break down in the soil, but not in the way that we want. It is changed – by microbes, and also by the aid of sunlight – into another compound usually known as photodieldrin. In this form it is remarkably resistant to biodegradation. This altered version of the molecule can persist in the soil for years. When considering how to produce chemicals for the future, we need to look into the way that a compound will disappear, as well as the way it behaves in the short term. Fortunately, there is a bacterium of the *Pseudomonas* genus that can contain the genes that enable it to degrade 2,4,5-T in the laboratory, and an *Actinobacter* that contains genes for PCBs. In the future, we may be able to harness these microbes to help reduce the poisons that we have put into our environment.

There is one crucial area of intervention for which we owe an ancient debt to our microbial cousins. I have left it to the end of the chapter, in the hope that readers will either have decided to skip to the next section on new diseases, or be swept along by a sense of momentum and digest this section along with the rest. The topic I have in mind is sewage. There, I said it. Just picture your local township at eight in the morning. Thousands of people sitting on thousands of toilets. Thousands of flushes working at about the same time, and thousands of specimens of faeces heading off downstream, all obediently heading – where? It is a question that nobody asks. The most profligate consumers of resources are Americans, and the American way of life provides an exemplar of the hunger that drives civilized society. Each day an American is powered by 8.5 kilograms (19 pounds) of fossil fuel, consuming 1.8 kilograms (four pounds) of food and 550 kilograms (150 gallons) of water (much of it used by washing machines and dish-washers). That is what goes into an American home. Out of it comes 0.9 kilograms (two pounds) of pollutant gases, 1.8 kilograms (four pounds) of rubbish and 450 litres (120 gallons) of sewage. The sewage contains less than two per cent organic waste matter, much of it coming from waste-disposal units that grind kitchen waste into tiny particles; the rest is water. The water is needed again, of course, and the job of the sewage treatment plant is to reclaim the water for someone else to use.

It has been said that, by the time a Londoner drinks a glass of water, it has already passed through nine previous persons. None of this is in the least distasteful. Excreta has an interesting composition, being one-third composed of the microbe *Escherichia coli*. This organism has received a bad press recently, and the notorious type known as *E. coli* O157 is dealt with in the next chapter. But don't blame the entire population because of the errant behaviour of just one miscreant – most strains of *E. coli* are not only harmless, but they are a helpful inhabitant of our bodies. The bacteria that live in our intestines are responsible for helping us make vitamins and they take a part in digestion. *E. coli* is important in another way, too: because it lives in the gut, it can be used as an indicator of sewage pollution. If you can find in a water course the type of bacteria that live only in humans, then you can safely conclude that the water has been polluted by sewage – that way you do not have to go round sniffing for faeces. *E. coli* is a friend and ally, not our foe. If there are pathogenic organisms in excreta, the contamination of water or food with those germs can spread a disease far and wide. Epidemics of such major diseases as cholera, typhoid (and, more recently, *E. coli* O157) are spread by such means, so it is important that sewage is made completely safe before it is disposed of.

As oxygen is needed to break organic matter down into water and carbon dioxide, we can calibrate sewage wastes by working out how much oxygen it takes for the organic content to be completely broken down. River water requires no more than about 100 milligrams of oxygen per litre (10 ounces per gallon)for all the organic matter to be oxidized, whereas a rich nutrient soup would require 70 or 80 times as much. This gives the biological oxygen demand (BOD) and this has become the standard way to describe the amount of organic pollution in a water sample. The primary treatment of sewage involves running it into settling tanks, so that the larger components are left at the bottom. These include pebbles and glass, dropped coins, condoms and jewellery. After about two hours, the liquid that remains can then be given secondary treatment in an activated sludge plant. The liquid is vigorously stirred, and oxygen is bubbled through it. The operators have to do nothing else, because the aerobic microbes naturally present in the system quickly metabolize the organic material and the BOD goes quickly down as the process proceeds. Some research has been done into the communities of microbes that take part in the process, and large numbers of different bacteria and yeasts have so far been identified. These are the organisms that normally break down organic residues in a river (including dead sheep, for instance) but in these activated sludge tanks they have a field day and grow at a great rate. The bacteria are themselves consumed by ciliates and other small organisms, and the BOD of the resulting water is reduced by 95 per cent. In some plants, the sewage is digested without pumping air through the tanks. These anaerobic digesters do not have enough oxygen to produce carbon dioxide and water; the end-product gas is methane instead. These plants often use the methane as the fuel to run themselves. Anaerobic digesters can handle concentrations of contaminants 20 times as great as an activated sludge plant, but they reduce BOD by only 75 per cent. Plant operators apply the most appropriate method for the sewage that they have to purify. A third approach to the problem, much used in sunny areas, is to run sewage into treatment lakes. Algae grow in the sunlight, and they serve to produce the oxygen that aerobic bacteria need to oxidize the wastes. So fast do they grow that they are removed from the water by the ton. These algal microbes are rich in protein, and can be processed to produce cattle feed. They have to be blended with other foodstuffs, because they are too rich in protein to act as a sole source of food. There is enough of this microbe protein produced by sewage processing in the USA to feed a quarter of the cattle population.

Some sewage plants then use tertiary treatment, in which water is run into ponds, or through tall towers, to cleanse the water finally. In any event, the water is finally filtered and treated with a substance like chlorine to remove germs. The levels of chlorine are specially controlled, because there should be no residual chlorine in tap water. During tertiary treatment, nitrates can be reduced by bacteria such as *Pseudomonas,* which break down the nitrate to release nitrogen back to the air. However, it is not only germs and nitrates that pose a hazard from sewage. In industrial areas there are sizeable amounts of chemical pollutants too. Indeed, this is one of the reasons why sewage sludge is not more widely used as a fertilizer, which would seem (at first sight) to be a viable proposal. It is a salutary thought that urban sewage sludge is too heavily polluted with heavy metals to be of use on the land. Treated sewage in rural areas is virtually free of metallic elements. Cadmium and chromium are usually absent altogether, though you might find a few parts per million (p.p.m.) of lead and perhaps 0.75p.p.m. of nickel. Compare that with the sewage sludge from an industrial town. Here the picture is dramatically different. You can find more than 100p.p.m. cadmium and nickel, and over 500p.p.m. chromium and lead. In rural sewage sludge there may be 100p.p.m. zinc, which rises to over 2,000p.p.m. in the cities. We need to take far greater precautions to prevent this wholesale pollution of the environment. Not only are such levels too high to risk putting the sludge on farmland (where it could be a valuable and natural fertilizer), but these metals can poison the microbes that we need to purify our wastes, and interfere with efficient sewage treatment.

Disposing safely of sewage has been perfected by humans over many centuries. Many of the traditional techniques we now use (like the methods we use to process food and produce leather) have been perfected without anyone realizing the important part played by the microbe. Most surprisingly, very few people comprehend that it is the microbe that makes the system work. This relationship with our microbe neighbours is intimate and has been going on for a long time. We tend to think of microorganisms as bent on causing us harm, but that is a distortion: the bulk of the microbial world is protecting us, serving us, helping to control pollution (and doing a far better job of that than humankind has ever managed) and regulating our environment. The microbes that purify our drinking water are an excellent group on which to end this chapter. The idea of drinking water that has already passed through a dozen people's bodies may seem distasteful, but (thanks to the microbe) the water that we obtain from a treatment plant is a lot purer than what we started with.

6

New diseases

SOME MICROBES ATTACK HUMANS. Humans have long been haunted by germs that cause terrifying diseases – such as cholera, typhoid, meningitis and smallpox – and we see them as a rival world that threatens our future. Over the centuries we have progressively hit back. Typhoid and cholera are rarities in the Western World, and bacterial meningitis can be cured if the treatment starts sufficiently early. Smallpox has been eliminated entirely.

In recent years our triumphant campaign has suddenly been thrown into reverse. For the first time in human history we are faced with a raft of new germs. If you look at a medical textbook of a generation ago, many of these were not even known. Those that were recorded occurred as rarities, whereas we now encounter them in the daily newspapers. We have gone from conquering traditional diseases to liberating new ones: *Chlamydia* and *Campylobacter*, necrotizing fasciitis and *E. coli* O157, BSE (bovine spongiform encephalopathy) and *Listeria*. Their ability to strike terror into our hearts, and to infect us apparently indiscriminately, gives them the status of highly evolved agents of disease, as though intent on human destruction. That is not a reasonable reflection of reality. Disease germs are maverick organisms, and are likely to harm themselves as much as their host. The truly successful germ is one that coexists with its host species, causing little actual illness. In this sense, the virus that causes the common cold is far more successful than the smallpox virus. The coryza virus, as the germ of the common cold is known, lives in everyone's nose. It disguises itself to avoid being identified as a dangerous outsider, and its genes allow it to change from one type to another every time a new antibody appears to tackle it. This gives it a uniquely intimate relationship with ourselves, and the most we know is the occasional cold.

We should begin with a celebration, so let us consider first the story of smallpox. Many former diseases have been brought under control in the Western World, but smallpox was the first great epidemic disease to be totally eliminated. The virus of variola, smallpox, produced a terrible scourge of humanity which was likely to kill and which left survivors scarred for life. For centuries people had practised variolation, in which pus from the victim of a relatively mild form of the disease was scratched into the skin of a young adult. These patients developed smallpox (hopefully of a similarly non-lethal nature) which – if they survived – would confer a life-long immunity. An ancient tale was that milkmaids were uniquely able to resist the disease. There are nursery rhymes which speak of the fair-skinned milkmaid, and the belief grew up that they contracted cowpox from the cattle in their charge, and that this disease conferred an immunity to the deadly smallpox. Folklore repeatedly substantiated this view, and some people tried to catch cowpox deliberately in order to obtain protection from its more deadly counterpart. Edward Jenner, country doctor and naturalist, was the person who proved the association and became the first person to institute a means of conquering an infectious disease.

Jenner was born at Berkeley, in the rolling English county of Gloucestershire, the eighth child of the Reverend Stephen Jenner, but he was left an orphan as a child. By the age of 13, he had left school and was apprenticed to a surgeon named Daniel Ludlow. The young Edward was himself the subject of variolation. He was deliberately infected with a mild case of smallpox, after several weeks of preparation by going on a starvation diet and being repeatedly bled. He was inoculated with the smallpox pus, and left in the 'smallpox stables' to suffer in isolation. In 1768 there was an outbreak of cowpox among the milkmaids – and as time went by, he noted that it was these very people who remained free of smallpox as successive outbreaks took place. The greatest single stimulus to Jenner's scientific work came from Captain Cook's expeditions to the South Seas. Joseph Banks was the great botanist on those voyages and, when he returned to England from Cook's expeditions in 1771, he approached John Hunter (1728–95), the surgeon, experimentalist and teacher, for advice on an assistant. Hunter had taken on Jenner as an assistant, and he recommended him for the job of sorting out and classifying Banks' extensive collections of new Australian plants. Jenner's work was so diligent that he was offered the post of botanist on a later expedition on board the *Resolution*. He declined the offer. Jenner loved the life of an English country gentleman, and had no wish to travel round the world.

Jenner was always interested in country life. As a child he spent much time collecting fossils and the nests of field mice. Hunter kept up an unceasing correspondence with him on matters of natural history: 'Have you caves where Batts go to at night?,' he wrote to Jenner. 'Are there no batts in the castle at Berkeley? If you have I will put you upon a set of experiments concerning the heat of them at difft seasons.' In another letter he asked: 'Have you got the bones yet of a large porpass? Is ever salmon spawn seen, if it is I wish you could get some.' At Hunter's insistence, Jenner carried out experiments on the hibernation of hedgehogs. When the Montgolfier brothers made their balloon flight in Lyons in 1783 – followed by an equally daring flight by Charles in Paris – Jenner made a balloon of oiled silk, filled it with hydrogen, and launched it on a successful flight the following year.

Edward Jenner was introduced to the Royal Society, but not for his work on smallpox inoculation. He had studied bird life, and on 13 March 1788 he announced that he had observed a young cuckoo ejecting the eggs of nestlings from the nest of its host. He went further, describing the hollow in the back of the young bird which enabled it to lift the eggs of its foster mother out of the nest. He went on to explain that the young cuckoo needed 15 weeks to come to independence, whereas adult cuckoos were only resident in Britain for 11 weeks. Surely, he argued, this was why the bird opted for a parasitic life – it was necessary for the survival of the young. The Royal Society was impressed by his logic and his diligence, and he was elected to the Fellowship on 26 February 1789.

As a pathologist, he wrote vivid accounts of his observations. He described calcification of the coronary arteries in his bold and casual style: 'When making a section of the heart, the knife struck against something so hard and gritty that I well remember looking up at the ceiling conceiving that some plaster had fallen down, but on further scrutiny the real cause appeared. The coronary arteries had become bony canals.' All the time Jenner was aware of the controversies over inoculation as a means of warding off the worst excesses of smallpox. Inoculation of young people with the virus of mild cases of smallpox had been formally established in 1701 by

Giacomo Pylarini in Constantinople. By 1746 Bishop Isaac Maddox of Worcester had set up institutions for inoculation and was widely preaching its value. Many noted doctors agreed, including Richard Mead in 1750, Robert and Daniel Sutton (1760 and 1767), Thomas Disdale (1767) and Théodore Tronchin (1770). It was introduced in Austria by the famous naturalist Jan Ingen-Housz (1730–99), in the face of fierce opposition. What preoccupied Jenner was the alternative approach: rather than inoculation with mild smallpox virus, he knew that victims of cowpox seemed to be permanently immune to smallpox encountered later in life. By 1790 he first inoculated volunteer milkmaids with pus from cases of smallpox, first using inoculations from mild cases and then moving on to samples taken from patients who had died of the most virulent form of the infection. In each case the result was unmistakable – the milkmaids were indeed completely immune to the terrible disease. They had already suffered from vaccinia (cowpox) and the protective value of this disease was unmistakable.

In Jenner's mind formed the idea of vaccination. If variolation was the inoculation of pus from a victim of variola (smallpox), vaccination would be administration of a dose from the victim of vaccinia (cowpox). The idea was already widely spoken of in the Middle East, and even in Britain Jenner was not the first person to carry out vaccination. This crucial experiment was first carried out by a Dorset farmer, Benjamin Jesty (1737–1816). He vaccinated his two sons and his wife with cowpox as early as 1774, and they remained free of smallpox from that date. Edward Jenner's experiment did not take place until 14 May 1796. That day, he inoculated a young boy named James Phipps with pus from a cowpox sore on the hand of a milkmaid named Sarah Nelmes. James became slightly feverish for a few days, but quickly recovered. To this extent, Jenner's experiment was no improvement on the pioneering vaccination by Jesty 22 years earlier. Two months later, that changed. Jenner took some pus from a patient dying of smallpox and scratched that into the arm of James Phipps. The brave young volunteer remained perfectly healthy. As Jenner wrote: 'The boy has since been inoculated for the smallpox which as I ventured to predict produced no effect. I shall now pursue my experiments with redoubled ardour.' He went on to publish *The inquiry into the causes and the effect of variolae vaccinae, known as smallpox* in 1798. The book became an instant best seller, and a specially bound copy in crimson velvet was presented to King George III. Although vaccination became socially fashionable, it attracted many detractors. Best known of these was the cartoonist Gillray, who drew deformed victims of vaccination as half-human, half-cow. We do not have to look far to see similar reactions against some of today's scientific innovations.

Jenner's next publication was *The Origin of the Vaccine Inoculation* in 1801. Here he predicted that vaccination would bring an end to smallpox: 'It now becomes too manifest to admit of controversy that the annihilation of the smallpox, the most dreadful scourge of the human species, must be the final result of this practice.' This was a brave prediction, because between 200,000 and 600,000 people were dying of the disease across Europe every year and a third of deaths in childhood were from smallpox. This disease brought down the House of Stuart, ended the civilization of the Aztecs and the great culture of the Incas when explorers introduced it to America. In 1779, it disrupted the Franco-Spanish Armada, and in 1870 decided the outcome of the Franco-Prussian War. There have been major epidemics in living memory, and the World Health Organization was prescient in establishing a plan finally to rid the world of the disease. At an historic meeting on 8 May 1980, the thirty-third World Health Assembly agreed a resolution which declared that smallpox had finally been eradicated. It is the first and only human infection ever to be eliminated. This is because there is no animal reservoir, and a related disease – equine pox – has also been eliminated in the horse population. In today's world, the only place where the smallpox virus exists is in freeze-dried form in closely guarded laboratories in Russia and America. It has ceased its reign of terror, and smallpox the disease is believed to be forever extinct.

Where is smallpox virus today? The official view is that after the disease disappeared there were three nations that possessed collections of the virus: Russia, the USA and South Africa. South Africa was induced to give up her stocks in 1983, leaving the two super-powers as holders of the virus in high-security laboratories. The date of 30 June 1999 was agreed as the final date by which Russia and the USA would finally destroy their remaining stocks.

In truth it may not be so simple. There is an outside chance that the viable virus might remain in some of the dead bodies of the last victims to die. If so, they could be dug up and the virus resurrected at some time in the future. It is even possible that other countries have kept supplies for military purposes. Since the opening of the former Soviet Union, there have been reports from senior scientists that Russia amassed large numbers of missiles loaded with smallpox virus. Research into the uses of smallpox virus as an agent of warfare were conducted by many nations in the 1960s, but in 1972 this came to a halt. In that year an international treaty was signed. The Biological and Toxic Weapons Convention agreed to outlaw all future research into germ warfare, apart from defensive investigations.

The reports from former Soviet officials claim that the effect of the ban was to redouble the secret production of the virus. It is claimed that Russia produced weapons capable of carrying smallpox virus to many nations round the world. During the years of the 'cold war', many communist nations were given supplies of the virus by Russia, including North Korea and China; other nations – such as Israel and India – were said to have produced supplies, along with parts of north Africa and the Middle East. Libya, Syria, Iraq and Iran have all been alleged to hold stocks of their own. Evidence for these claims is slight, as you'd expect, and these are only rumours.

Meanwhile, it is a fact that one of the permitted exports to Iraq was growth medium for the culture of pathogenic bacteria. Growing bacteria in this way is an important part of medical practice. This is how we identify bacteria, and find out which antibiotics will be best to treat the patient. On the other hand, it is also the way you would grow huge amounts of deadly bacteria for use as possible agents of war. The facts are that huge amounts of growth medium were exported, and it is very hard to imagine how these amounts could be used by hospital laboratories.

Bacteria and viruses are very easy to produce, and the pathogenic types can be extraordinarily potent as agents of disease. The fears for biological agents of war remain, and it takes more than a surprise visit to be absolutely certain that a nation with hostile intent is not conducting this research in clandestine conditions. You could raise smallpox virus in a garage, without any sophisticated apparatus. Germ warfare is bad enough; if it becomes gene warfare we shall have unimaginable problems.

During the battle against natural smallpox epidemics we came to learn a great deal about the virus, and yet some mysteries remain. The fact that the genes of the cowpox virus could confer immunity to smallpox was a fortunate coincidence. As Jenner knew, there are several similar diseases of cows which he called spurious cowpox. We now know that some of these are caused by parapox viruses, while the apparently similar bovine mammillitis is caused by the herpes virus. Cowpox does survive in other species, and occurs in cows, cats, humans and wild rodents. Interestingly, the virus that was widely used in vaccination does not have exactly the same genes as cowpox. In 1939, it was shown to be a distinct virus and nobody knows where it originated. One theory is that it was a variety of equine pox. Even though variola is extinct, this vaccinia virus is still used in medical research. A version of the virus containing engineered genes has been used to protect against some other infections – it showed early promise as a vaccine against rabies, and was used to try to halt the spread of this terrible disease in the European fox population. Many scientists prefer the idea of using bird pox viruses (such as canary pox). These can also be engineered to produce vaccines, but – unlike vaccinia – they cannot be replicated in humans.

What other diseases could we eradicate? There have been hopes that the guinea-worm which causes dracunculiasis could be eliminated, and there ought to be a war on *Helicobacter pylori*. This bacterium has recently been discovered to cause peptic ulcers and gastric lymphoma, and to be associated with cancer of the stomach. The great pharmaceutical companies are energetically promoting drugs to hold these diseases in check – chemotherapeutic agents for the cancers, and antacid tablets for the ulcers (all of them highly profitable products) – and are far less keen to promote the idea that we could control many of these cases by eliminating a common bacterium. Already we have some simple tests to determine whether someone is infected with the bacterium. There is also a cocktail of antibiotics that removes *H. pylori* in most patients. If we could eliminate the bacteria from human society we would make tremendous inroads into the control of three major diseases. This organism has been found only in humans, and a campaign to develop a suitable vaccine to eliminate the bacterium could help overcome some current diseases of widespread significance. Poliomyelitis is one example of how we might succeed, given enough time and energy. This paralysing disease is now vanishingly rare across the northern hemisphere, although people crippled by its effects are a surviving reminder of the days when it was rife. Other diseases of childhood also controlled by immunization include mumps and measles. With time, most of these can be controlled and many actually eliminated, just like smallpox. We have almost certainly eliminated one serious epidemic disease, vesicular exanthema of swine, which was first documented in 1932. It was recorded as a disease similar to foot-and-mouth disease among pigs in California. Every infected animal was slaughtered, the farms quarantined and the area disinfected. Within months it had been contained. The next outbreak occurred in 1933, and research showed that, although related to foot-and-mouth, this was a new disease. It was then given the name vesicular exanthema of swine (VES). Regulations were introduced to oblige all those who kept pigs to cook the scraps they were fed. It proved difficult to police, and over the following 20 years outbreaks continued. Up to one-fifth of the entire pig population of southern California was affected, amounting to more than two and a half million animals.

By 1950 the disease had become rare. Then, in 1951, a garbage train left San Francisco with a load of uncooked pig scraps which were off-loaded in Wyoming. The food went to a single pig farm. When some of the animals became lame and unwell, the farmer quickly sold them on to farms all over the USA. From 1952 to 1956 outbreaks occurred in 40 of the 48 states. The laws were tightened. Penalties and policing were increased and many states made it illegal to feed scraps to pigs at all. The last case was in New Jersey in 1956. From that time on there has been no further outbreak, and the disease never escaped from the USA. It is now believed to be extinct, solely because of rigorous action and clear disclosure of the risks.

Yet suddenly we are faced by new diseases. Some are interesting new discoveries, others – like AIDS and the potentially cataclysmic outbreak of BSE – are major problems which can teach us crucial lessons for the future. Now is the time to set down what they are and what they do, because detailed knowledge of these new diseases can be hard to come by. Some of them are conditions that you may never hear about. There is a tiny amoeba that infects contact lenses and has caused severe damage to the sight of many people. This organism is *Acanthamoeba keratinoides* and it lives in the thin layer of fluid beneath the contact lens where it eats away at the cornea. The organisms can only survive where hygiene is poor, so it is important for wearers of these lenses to sterilize them regularly. It is unnatural to trap a layer of liquid on the surface of an organ like the eye, with a well-developed ability to cleanse itself, and which evolved to endure the atmosphere. Here we encounter an important first principle: when we adopt unnatural means to solve our problems, we must first learn lessons from nature. Many of the new diseases are spread because of new processes that we are introducing without thinking them through. Chilled food cabinets encourage the growth of some microbes, for example, and the mincing of meat allows bacteria (traditionally killed by cooking) to survive deep within a hamburger. Modern methods of rearing poultry have led to outbreaks of *Salmonella*, and current opinion suggests that BSE was sparked by the way we feed beef animals. The most surprising fact of all is the way that food poisoning is on the increase. For decades it has gone down, but in recent years the number of reported cases have been shooting up again. The number of unreported cases must make the total far greater. Did you report the last time you had food-borne diarrhoea to the authorities? Neither did I.

Food can spread a large number of diseases, partly because it is often served warm (an ideal temperature for most bacteria), and also because it is nutritious. If it will make a fine meal for you, it makes a perfect meal for the microbes that cause food poisoning. You may not have heard of *Bacillus* as a cause of food poisoning, but this genus contains several species that make people ill. The bacilli live in the soil and dust or among vegetation. When they are allowed to breed in food they can increase in numbers and make the consumer ill. The most familiar of these organisms is *Bacillus cereus*, which produces sickness, diarrhoea and abdominal pain. Other cases result from *B. subtilis*, which causes nausea and diarrhoea (but not much pain) or *B. licheniformis* which causes diarrhoea and abdominal pain (but not the nausea). These bacteria have been known for over a century, and there is nothing new about their behaviour. What is making them into a new threat is the way that we prepare food these days. All these bacilli form spores which can survive drying and heat, and the spores will hatch out if the conditions are right. The problem that gives rise to this form of food poisoning is when freshly cooked food – still contaminated by spores – is stored without being sufficiently cooled. If the foods are eaten when freshly cooked, there is no problem. If food is cooled and stored in the fridge, there is little risk either. But when the food is not properly cooled, so it spends hours still warm, the bacteria can grow rapidly and this is where the problems arise. Each species has its own preferences. *B. cereus* grows in rice dishes, and sometimes through pasta or sweet pastries. *B. subtilis* and *B. licheniformis* prefer pastry dishes made with meat, including ethnic meat and seafood dishes. Traditional foods were prepared and eaten promptly, but when modern society separates the production and the consumption of food, there is time for bacteria to grow in places where they were once absent.

There are many species of *Salmonella* sometimes found in the modern food supply. Their name commemorates Daniel Salmon (1850–1914), a leading American veterinary pathologist. These bacteria are all liable to infection with phage viruses (see page 142), and we keep track of the different strains of bacteria by observing which phage virus attacks them. *Salmonella* produces a disease with high temperature as well as diarrhoea and vomiting. The bacteria are found in animals, particularly poultry and some pets, including terrapins. Thorough cooking always kills *Salmonella* (it cannot form heat-resistant spores, like the bacilli) but food that is incompletely cooked can incubate large growths before the food is eaten. Eggs are sometimes contaminated by *Salmonella*, but if they are cooked until the yolk is hard then the bacteria are killed at the same time. Poultry are often contaminated with *Salmonella*, and we eat a lot of poultry. The annual mass of chickens consumed in Britain is over a million tonnes, adding up to 13 birds for each person. We are frequently told that there is little we can do about the contamination of poultry by *Salmonella*. This is a cover-up. If authorities decide to control this pathogenic organism, there are plenty of ways to tackle it. Sweden has a zero tolerance of *Salmonella* in poultry. Tests are carried out to trace the organism and it is ruthlessly controlled. As a result, Swedish chickens are *Salmonella* free. Everyone else's could be too.

Salmonella has been known about for a century, but there is a new threat from poultry – a bacterium named *Campylobacter*. This was unheard of a generation ago; the organism was only discovered in 1974. Now it is everywhere. A survey by microbiologists at Aberdeen University concluded that 100 per cent of chickens passing through a processing unit ended up infected with the organism. Not all of the birds were infected when they arrived at the processing plant, but every single one was infected by the time they came out. Infections with *Campylobacter* were rare in the 1980s; as the millennium draws to a close it has been estimated that there have been half a million cases of food poisoning caused by this germ in Britain every year. Interestingly, *Campylobacter* does not reproduce on food, but only in the living chicken. Most major outbreaks have been related to contaminated water supplies or inadequately treated milk. The main sign is gastroenteritis, with diarrhoea and vomiting accompanied by severe cramps and dehydration. One person in 15 develops a reactive arthritis, and in a few there are the symptoms of Guillain-Barré syndrome, a creeping paralysis. This is a kind of autoimmune disease, and it can even result in a need for patients to be maintained on a life-support machine if breathing fails. In most cases it clears up within a few weeks, but it is a dreadful condition to suffer while it lasts.

The exact extent of the problem is hard to determine. The overwhelming majority of cases of food poisoning are never reported, of course, and the industry always down-plays the extent of infection. The officials who reacted to the Aberdeen University report denied that 100 per cent of poultry carcasses were infected – but conceded that the total might be 40–90 per cent. A recent survey of 1,000 fresh chickens bought in stores across the USA showed that 63 per cent were contaminated by *Campylobacter*, and 16 per cent were infected by *Salmonella*; eight per cent proved positive for both. The perceived quality of the chickens was no guide to their safety, indeed the 'premium' selection (including free-range birds) were most likely to be infected.

For most of us, the best way to control the bacteria is to assume that every chicken you ever handle is dangerously contaminated – but to kill off all the germs by thorough cooking. Of course, your hands may have been contaminated by handling the bird before cooking it – so personal hygiene is very important. They always tell you to be sure to wash your hands before cooking food. I would be as keen to encourage people to wash their hands afterwards. We are always told to wash our hands after visiting the lavatory but (particularly for men) I believe it is just as important to wash your hands before you go. Your vital organs have been tucked up inside your underpants, but you never know what you might have touched meanwhile. You can always tell the microbiologists in a public lavatory; they wash their hands thoroughly before unzipping their pants, but rarely bother to wash afterwards.

We have heard of *E. coli*. It is important to remember that 'E-coli' is not correct. Scientific names are always printed in italics, and the initial of the first name is always a capital letter. You, dear reader, are *Homo sapiens*, and not *H-sapiens*; and it is *E. coli*, not *E-coli* as the *Sunday Times* likes to say. This species is ordinarily nothing to fear. Indeed, the bacteria live in our intestines in huge numbers, and they are normal inhabitants of the ecosystem inside the human body. The bacterium was named after Theodor Escherich (1857–1911) who first discovered it. He named it *Bacterium coli*, and it was subsequently renamed *Escherichia* in his honour. Most strains of *E. coli* are harmless, and many of them may be necessary for the healthy functioning of the gut. We can tell one strain from another through the phage viruses to which they are sensitive, and it is just a few of these phage types that cause serious problems. The genes of the strains of *E. coli* that inhabit our intestines fit them perfectly for their role, but some strains have acquired genes from other bacteria – and these genes can code for the production of virulent toxins. The organism known as *E. coli* O157 picked up a couple of toxin-producing genes from a bacterium called *Shigella* in the USA in the early 1980s, and began to spread around the world. In the early 1990s a small percentage of Western cattle were carriers; now most of them are. Curiously, it is not found in tropical countries and (even more strange) it does not make cattle ill, even though it can kill humans. It produces a widespread toxic disease of the body with blood in the urine and faeces and a breakdown of many essential organs (like the kidneys). It is spread in dishes like hamburgers, where the mincing process spreads bacteria on the surface throughout the meat to parts where the heat of cooking may not be sufficient to kill them. Traditional cookery of meat involves the searing of meat by heat or boiling it in a stew, both techniques killing any of these bacteria on the surface. It is the modern method of mincing food – carrying surface bacteria deep inside the finished product – that increases the risk of an *E. coli* O157 infection. Hamburgers are an excellent home for germs, and the exposure of employees to the large amounts of potentially contaminated beef gives plenty of opportunities for the germs to spread. One case resulted from a four-year-old little boy petting a goat in a zoo. In today's world, consumers are best advised to regard any sample of supermarket meat as a possible source of infection. Again, washing your hands after handling food is just as important as washing them before you cook.

The related organism *Shigella* is named after a distinguished Japanese microbiologist, Kiyochi Shiga (1871–1957). *Shigella* is the causative agent of dysentery, in which the genes code for the production of toxins that attack the lining of the intestines. The result is a bloody diarrhoea. The most common species, *Shigella sonnei*, is commonly found in Britain and America and produces a form of dysentery that is not severe, but other species (including *S. boydii* and *S. flexneri*) are brought in from other countries and in these cases the disease can be severe. Dysentery caused by *S. dysenteriae* is the worst of all, and can result in blood in the urine and toxic poisoning of the intestine. The disease is sometimes found in institutions, often associated with young children. Poor hygiene results in the organism spreading from the faeces of one victim to the mouth of a new contact, and in this way an outbreak is triggered.

The genus *Clostridium* contains a range of dangerous bacteria, most of which live in an oxygen-free environment. Most of the species have been known for a century. Tetanus, for example, comes into this group. *Clostridium botulinum* is the organism that produces botulin, one of the most poisonous toxins known to science. Although it does not come into our category of a new disease, it has recently been found in one unexpected situation. American mothers sometimes like to dip a comforter or dummy in a jar of honey, and *C. botulinum* is sometimes found in honey. The result was a low-grade level of poisoning which took diligence to solve. Sometimes, old bacteria turn up in new situations as human behaviour changes over the years. There is a new addition to the list of species in the form of *C. difficile*, which causes an infection of the intestines with severe inflammation and the production of fluid in the gut. This species is found in about three per cent of normal people and over ten per cent of hospital patients. Normally, it causes no problems, but if a patient is treated with antibiotics the bacteria can grow out of control. Almost all antibiotics are capable of triggering an infection with *C. difficile*, and the infection can spread from one person to another once it has become established. The bacteria produce toxins (enterotoxin A and cytotoxin B) which attack the body's cells and cause severe effects to the lower intestine.

These organisms all cause their own form of gastroenteritis, but there is one newcomer that is truly life-threatening and insidious. This is *Listeria*, a bacterium named after Joseph Lister (1827–1912), who was the first person ever to grow a bacterium in pure culture as long ago as 1878. The bacterium is found in the environment: in cattle and sheep, in silage and soil. In humans they produce a 'flu-like illness which can even mimic meningitis and progress to septicaemia. Most dangerous of all is the fact that *Listeria* can produce a lingering infection without symptoms. In pregnant women it can be particularly dangerous, because it can cause spontaneous abortion of the fetus, even when it produces few symptoms in the pregnant woman herself. *Listeria* has been found in a range of raw vegetables and processed foods. It has often been reported in soft cheeses and meat pâtés, and can grow slowly even in the fridge. This is an insidious organism with potentially grave effects, and pregnant women throughout the English-speaking world are regularly advised to avoid such foods.

Some of these organisms can occasionally be spread by contaminated water. *Shigella* and *Campylobacter* have both been spread in this way, and some of the new diseases are exclusively spread by water. *Giardia* is a pretty little cell, pear-shaped and concave, rather like a leaf. It swims with eight long flagella so it flutters gracefully along. In the human host it attaches itself to the wall of the intestines by the sucker on its concave side, and produces an intestinal disease with diarrhoea, known as giardiasis. *Cryptosporidium* is another water-borne organism. It has long existed in farm animals and pets, but now occurs as an occasional contaminant of

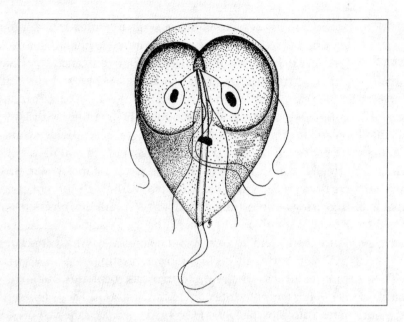

A face looks up at the observing microscopist. With its paired nuclei looking uncannily like eyes, a cell of *Giardia* is well known to microbiologists. This organism is frequently found in beech-wood ponds, where it can be seen travelling with a flickering motion, propelled by its trailing flagella. Some species cause intestinal disease when contracted from drinking water.

drinking water. Neither *Giardia* nor *Cryptosporidium* is affected by chlorination, and both produce unpleasant inflammation of the intestinal tract. The chances are that earlier generations were regularly exposed to *Cryptosporidium* by living on close contact with farm animals. They would have caught the germ and suffered a bout of diarrhoea – but after that they would have a life-long immunity to the disease. Most people in modern society live far from farms, and grow up without the immunity. A chance infection with the bacterium causes a rapid and unpleasant diarrhoea, and in some groups of people (elderly people, for instance) the consequences can be life threatening. In recent years there have been many outbreaks spread through drinking water supplies. More than half a million people in south-east England were told to boil their water after one outbreak, for example, and the economic consequences could prove to be serious. Although *Cryptosporidium* is not destroyed by conventional chlorination, it is effectively controlled by slow sand filtration (which, as we saw on page 75, relies on microbes to remove the bacteria). Modern water companies regard slow sand filtration as too slow, and have opted for faster – but less thorough – methods of treating their supplies. One day soon there will be a major outbreak, and the long-suffering victims will seek legal redress from the company that sold the contaminated water.

One of the most unusual new infections is Legionnaires' disease. It was first recognized in 1976, when 200 delegates fell ill after a meeting in Philadelphia. The victims of this previously unrecorded infection were members of the American Legion, and they became ill with respiratory disease, high fever, chills, muscle aches, and changes in behaviour and liver function. Some of them quickly worsened and died. Most of those who survived were treated with erythromycin, although nobody knew what had caused the outbreak of Legionnaires' disease at the time. Research showed that the infection was caused by a strange organism later named *Legionella pneumophila*. This germ normally infects amoebae living in water, and I have speculated that perhaps it grew just as well within the white cells of the victims' blood. White blood cells are, after all, amoebae with an independent existence and it would be a small change for *Legionella* to move from a free-living amoeba to one that is part of the human body. What I find interesting is the fact that the amoebae in which the organism lives are found in warm water, a good comparison with the confines of the internal human environment. For this reason, the hot water supply in many hospitals and other institutions is now kept dangerously hot (there are notices to warn the unsuspecting user). Hot water can kill the amoebae in which *Legionella* likes to live.

The original Philadelphia outbreak seems to have resulted from contaminated water in the air-conditioning being spread like an aerosol through the convention rooms. Great care is now taken by hotel keepers to prevent a recurrence, and hot water is a symptom of their determination to banish *Legionella*. The disease remains a problem. A scare in 1998 centred on the cruise ship *Edinburgh Castle*, which had reported symptoms among her 800 passengers and 400 crew. There are still 300 reported cases a year in Britain and the mortality is steady at ten per cent.

Species of *Chlamydia* produce a variety of diseases, chlamydiasis being one of the many infections spread by sexual contact. The infection causes few symptoms, but it can infect the cervix or the fallopian tubes. In many cases sterility results, yet the sufferer suspects nothing. One of the side effects of the widespread use of the Pill means that women feel confidently able to share sexual partners without the risk of pregnancy. Teenagers are sometimes prescribed the Pill in the belief that it keeps them 'safe', and manuals of sexual behaviour say little about sexually transmitted diseases. Sadly, there is a rising incidence of sexual diseases that cause disease in later life. Apart from chlamydiasis, we are faced with rising levels of cytomegalovirus, the papilloma virus, which causes genital warts, genital herpes, trichomoniasis and infections of the urethra caused by a range of conditions. Although these diseases are well known to microbiologists, they are less frequently discussed by the public. New strains are being discovered, and in 1980 a new species of *Chlamydia* was recognized. This organism, *C. pneumoniae*, was found to cause many cases of cough and sore throat, raised temperature, and – as the name suggests – sometimes a form of pneumonia. More recently, pathologists have even isolated *C. pneumoniae* from the fatty plaques which block the coronary blood vessels in heart attack patients. It may even be that the coronary heart attack, scourge of the modern manager, is caused by a germ rather than by stress or a dangerous diet. We already have one circumstantial line of evidence that supports this theory: the incidence of heart attacks seems to have gone down in recent years, and this may be because the widespread use of antibiotics has been removing the *Chlamydia* before it can cause any long-term damage.

There remain some mysterious diseases that seem to be caused by germs, but the cause has never been found. In 1967, a Japanese scientist, Tomasaku Kawasaki, first described the syndrome which now bears his name. It was originally described as mucocutaneous lymph node syndrome, and it is the main cause of heart disease in the young. It mainly affects children aged four or under, yet a quarter of the patients develop a hidden weakness of the arteries. The full effects of the disease do not appear for a decade or more. The disease attacks the blood vessels around the heart, and teenagers have been known to collapse as a major vessel bursts. The incidence varies from nation to nation: in Britain the disease is found in fewer than four cases per 100,000 children aged under five years, but in Japan the rate is more than ten times greater. There is still no test for the disease, and no firm conclusion over the cause. Research is urgently needed, and so is a greater public awareness of the threat.

We can be reasonably confident that it is the result of an infection, even though the germ has not been found. It has its peak incidence in children aged nine to eleven months, and 80 per cent of all cases are in the under four year olds. This is typical of other childhood infections. We have also observed epidemics, clusters, and a seasonal rise and fall in incidence, like many bacterial diseases. It isn't highly infectious, because it rarely seems to spread directly from person to person. It seems that this may be one of the many childhood infections we all face, but which produces its effects in a proportion of its victims – and some of these will later die. There have been many theories about the cause of the disease, including bacteria spread by mites, several well-known viruses including retroviruses, parvovirus and the Epstein–Barr virus which causes glandular fever, but none of these explanations stands up to scientific investigation and the cause remains a mystery.

There are some diseases caused by toxins produced by familiar bacteria such as staphylococci and streptococci, and known collectively as toxic shock syndromes. Kawasaki disease is remarkably similar to one of these diseases, and may yet prove to be an abnormal response to familiar bacteria. All these diseases show a sudden onset of fever with enlarged lymph nodes and a rash which sheds skin from the surface. The lips and tongue are often reddened and sore, and there may be conjunctivitis as well. Although we have no cure, there are treatments. One useful drug is aspirin, which seems to control the symptoms and may even help prevent long-term effects. The best treatment is the intravenous injection of gamma-globulin, a concentration of antibodies extracted from donated blood. If this is administered within the first ten days of infection, the disease is greatly dampened down and its consequences reduced. The widespread nature of Kawasaki disease has not been generally recognized, and parents need to know the symptoms so that prompt action can be taken. The typical diagnostic criterion is a high fever lasting for five days, and the presence of four of the following five signs:

1. Severe conjunctivitis
2. Reddening and soreness of mouth and tongue
3. Swelling, redness and flaking of the hands and feet
4. A generalized rash
5. Swelling and soreness of the lymph glands in the neck.

This is enough for parents to consult a doctor with a strong case for urgent investigation. Time is of the essence, and with understanding of the condition parents can be reassured that every case of a sore throat is not going to lead to a life-threatening disease.

Which brings us to meningitis, an illness that every parent dreads. Each time a child develops a headache, the parents fear the worst and, each time an outbreak of meningitis is reported in the press, the doctors know that they will be overwhelmed by anxious 'phone calls. Meningitis is an infection of the meninges, which are the membranes around the brain and lining the cavity in which it lies. The infection can be produced by several different bacteria and viruses, but the classic bacterial meningitis which has recently emerged in epidemics among the young is caused by *Neisseria meningitidis*, an unusual bacterium that occurs in pairs. The genus is named after Albert Neisser (1855–1938), the German bacteriologist and dermatologist famous for his work on sexually transmitted diseases. *Neisseria* is found in the human population, where it exists harmlessly in the nasal passages of healthy people. Once in a while it manages to gain access to the meninges, where it produces a rapid and severe infection of the fluid that bathes the brain. Occasionally the bacteria invade the bloodstream, causing meningococcal septicaemia. The effects are serious and urgent action is necessary. Having said that, the bacteria have not shown much increase in resistance to antibiotics, and penicillin is enough to eliminate them in most cases.

The problem is to get the penicillin into contact with the bacteria. The organisms are well protected inside the meninges, basking in the cerebrospinal fluid around the brain and down the spine. An injection of penicillin, which will work well enough against bacteria circulating in the bloodstream, does not rapidly reach within this sealed area. Meanwhile, meningitis is an excruciatingly painful condition, where waves of unendurable pain of increasing intensity scour the body – and keep getting worse. I went through it and longed, for a short time, for the relief of death, but recognized, even in a comatose state, that life was too promising to lose hold. The medics were convinced I was not long for this world, and told my relatives that the best thing to do was to pray. It was the most excruciatingly painful experience and – as life-threatening experiences always do – it further heightens one's zest for living. The diagnosis of the disease in the hospital is made by removing a sample of the cerebrospinal fluid and looking at the organisms under a microscope. The fluid in a healthy person is crystal clear; in bacterial meningitis it is as thick and cloudy as undiluted lemon barley water. My proposal, after this experience, was that doctors should replace the volume of cerebrospinal fluid removed for the test with an equal volume of penicillin in saline. This would have two effects: it would neutralize the hydrostatic balance (withdrawing spinal fluid can cause agonizing pain), and it would put a dose of the antibiotic exactly where it can do most good. This would be done before the identification of the organism – the treatment could do no harm if *N. meningitidis* was not found, and could be life saving if it was.

There is one intriguing indicator for inflammation of the meninges which I learnt from all this, and it is known as Kernig's sign. If a patient cannot straighten the lower legs while seated on a chair, as if raising them to point straight ahead, then meningitis or a closely related condition is indicated. It works as well if lying on one side, with the legs drawn up: the lower legs simply will not straighten at the knee. The ham-strings behind the knees become tightened, and prevent the leg extending forwards from the knee joint. If this sign appears, the urgent attention of a doctor is always required. The most popular 'sign of meningitis' is blotches on the skin (which do not disappear even if pressed with the side of a glass). However, this only applies to septicaemia, when the bacteria are growing in the blood. Normal meningitis does not behave in this way, and for these cases Kernig's sign can be crucial.

Meningitis in the past was caused by a great range of viruses and bacteria, some of which are now much rarer in Western society. It has been caused by the organism of tuberculosis, and even typhoid bacteria. In previous generations, death was always a threat. These days, with prompt medical treatment, recovery is frequent.

Mention of tuberculosis (TB) brings us to one of the diseases that had largely disappeared from many countries, but which is suddenly on the increase. The organism causing this disease is *Mycobacterium tuberculosis*, a bacterium with a waxy coating which makes it harder for the body to detect and attack. It was discovered in 1882 by the German country doctor Robert Koch (1843–1910). TB was always a common disease of childhood – most of the children contracting it never became ill. Instead, the bacteria would be walled up within a calcified nodule which showed up clearly on X-rays. The patient responded by producing antibodies to the bacteria, and these persisted for life. Their presence conferred immunity to the disease thereafter, because the antibodies were lying in wait and quickly inactivated any later invasion by the bacteria.

This offers a possible route to immunization against the disease, and the search for a harmless form of the bacterium which could still stimulate the patient's immune system to produce antibodies was started in the nineteenth century. The research was spearheaded by two French microbiologists: Albert Calmette (1863–1933) and Camille Guérin (1872–1961). Calmette was taught by Pasteur and had founded the Pasteur Institute of Saigon, where he discovered a serum to protect against snake venom. Back in France he founded the Pasteur Institute of Lille, and began working with Guérin on the problem of tuberculosis. By 1906, Guérin had shown that people who had been infected with tuberculosis when young retained a resistance to an infection later in life, and in 1908 they found that a virulent strain of *Mycobacterium tuberculosis* became transformed to a relatively harmless strain if cultured on a medium containing bile. They produced strains of decreasing virulence, until by 1921 they were confident that the organism they had obtained was harmless to humans – but would still trigger the immune response that gave the life-long immunity. This strain was used to make the BCG vaccine (named after the initials: Bacille Calmette Guérin), which they first used to vaccinate babies at the Charité Hospital in Paris in 1922. From then on, mass vaccination programmes using the BCG vaccine were set up in many countries – China and Japan, Russia and Canada, as well as in France. It worked well, until a major set-back in Germany. During a vaccination programme in Lübeck in the spring of 1931, 249 infants were given the BCG injections. This batch was contaminated with a virulent form of the bacterium, and by the autumn of that year 73 of the infants had died of TB. This set-back remained in the minds of many, and the international use of BCG was slow to gain acceptance. Although the vaccine was tried in England, it was stated during the 1940s that the use of BCG vaccine in Britain 'had not been met with favour'. American research into its effects resulted in it being progressively introduced about 1950, but the British Medical Research Council did not finally report on the value of the vaccine until 1956. Since that time, the routine use of BCG in Britain and America has slashed the incidence of the disease and indeed it often seemed to be close to extinction in the Western World.

The vaccine is also being used to stimulate the body's immune system in the battle against cancer. This technique works only for some tumours, because they have to be surface growths that bacteria can easily access. The technique is being used for the treatment of bladder cancer with much success. The treatment is the simplest that one could imagine: a suspension of the BCG bacteria is introduced into the bladder through a catheter and, after an hour, the liquid is let out again. That is all the treatment entails. The organisms are taken up by the cancerous cells, but then the patient's white cells detect the presence of BCG and set out to destroy the cells containing them. Within a few days, the patient notices the passing of small pieces of tissue in the urine. These are all that remain of the tumour. Cure rates of around 80 per cent have been claimed from this technique, which works simply by stimulating the patient's immune response and involves no chemotherapeutic medication. Interestingly, the use of BCG therapy for cancer parallels the introduction of the vaccine: the technique is widely used in the USA and Europe, but it has hardly been used in Britain. Enthusiastic doctors are claiming that this technique offers the most effective treatment for bladder cancer yet discovered, and it works by using microbes to stimulate our own cells to work on our behalf.

Tuberculosis is proving to be a more difficult problem, however. In a previous era, if an unprotected adult presented with the disease, there were many drugs available to eradicate the bacteria. Although penicillin was not effective, the next antibiotic to be discovered, streptomycin, had a powerful action against the bacterium. Alongside this drug were isoniazid and para-aminobenzoic acid (PABA). The tuberculosis organism has acquired genes that confer resistance, so other drugs have been brought into use, including ethambutol, rifampicin, thiacetazone and pyrazinamide. Meanwhile, the bacteria acquire new forms of resistance, and some recently identified strains are resistant to all available drugs. Some patients have died in recent years, beyond the ability of modern medicine to offer a cure. Others have been cured only through the surgical removal of an infected lobe of the lung, an operation not needed for decades. As a result, there is renewed interest in immunization and the risk posed by TB – which had vanished into insignificance for many of us a generation ago – has dramatically returned.

A similar story can be told about one of the most common inhabitants of the human body, *Staphylococcus aureus*. Most times, these bacteria live on us and cause no problems. Occasional bacteria enter through a hair follicle and produce spots or boils. Sometimes a strain with genes that confer virulence gets into the bloodstream, and causes a generalized septicaemia which can be life threatening. In America 70,000 people still die from *S. aureus* every year. Staphylococci are among the bacteria that are themselves attacked by a range of phage viruses, and here too we can identify the genetic strain by reference to the phages to which the bacteria react. Hospitals monitor their staphylococci, and are ready to tackle an outbreak if one of the particularly virulent phage types is detected. *Staphylococcus* was the first bacterium to be tackled by antibiotics, and penicillin was used to destroy these bacteria in many otherwise fatal cases. The first use of penicillin was in 1932, when a hospital doctor dripped broth in which the fungus *Penicillium* had been grown into the eyes of newborn babies whose eyes were infected with the gonorrhoea bacterium. The broth contained enough penicillin to eliminate the bacterium in these cases of ophthalmia neonatorum. Little babies, otherwise doomed to blindness, were cured as a result. As World War II loomed closer, increased interest in the prevention of infection focused attention on antibiotics and penicillin, which until then had been little more than a laboratory curiosity, went into mass production. The successful response in the first case to be treated – the policeman with septicaemia – taught a lesson to the doctors: penicillin could be used to treat a generalized infection, and supplies would be of the greatest value in a theatre of war. From the 1940s on, penicillin was the drug of first choice in staphylococcal infections.

Bacteria are great opportunists, however, and they can acquire genes from other strains. Just as with the organisms of tuberculosis, some staphylococci began to acquire genes, conferring resistance to antibiotics. Bacteria that could produce penicillinase soon appeared. This is an enzyme that breaks down the penicillin and prevents it damaging bacterial cells. In recent years, the position has worsened considerably, because strains of *S. aureus* that have multiple resistance have begun to spread. It has acquired new genes that make its cell walls resistant to penetration by antibiotics. The standby drug against staphylococcal infections has, for many years, been methicillin, a modified form of the penicillin molecule. These new bacterial strains are resistant to the antibiotic (methicillin-resistant *S. aureus* is known as MRSA for short) and different drugs have to be brought into use.

The most powerful anti-staphylococcal antibiotic is the rarely used vancomycin and hospitals now hold stocks of this antibiotic to tackle outbreaks of MRSA. The organisms are beginning to occur in the general population, but are far more common in hospitals. About 30 per cent of the *S. aureus* in hospitals are resistant to methicillin, and this figure rises to 40 per cent in intensive care units. For many of these patients, vancomycin is the only antibiotic available to treat their infections. A more disturbing recent development was the announcement, in the spring of 1997, of the isolation of a new strain of *S. aureus* in a four-year-old boy in Japan which had acquired genes that made it completely resistant to vancomycin. This is a most serious development: for doctors in the Western World, vancomycin is a drug of last resort. The antibiotic used in Japan to treat this little boy was arbekacin, a rare drug not licensed in the West. Most worrying of all is a new strain of *Pseudomonas aeruginosa*, which is often the cause of hospital infections in burns victims and patients with lowered resistance. A recently identified strain is unaffected by any of the antibiotics available for test. To date it is not known to have caused an infection in anybody, but an outbreak can only be a matter of time. For all the excitement about the reckless pace of medical research, we have seen no new antibiotics introduced into general practice for over 20 years.

It is often said that bacteria become resistant to antibiotics through a form of gradual selection. People say that the use of antibiotics allows the hardiest germs to survive, and (as they give rise to succeeding generations) a gradual increase in resistance takes place. This is a model that is easy to understand, but it isn't really what happens. Bacteria like to have sex with each other, and they acquire new characteristics by exchanging genes from time to time. This is how *E. coli* O157 acquired its ability to make lethal poisons. Bacteria develop resistance by this process, and not just through some mechanistic method of selective killing by antibiotics. Life is far more sophisticated than mechanics. The over-use of antibiotics is unlikely to be the main cause of resistance, because the dangerous strains of antibiotic-resistant tuberculosis bacteria arrive in the West from underdeveloped nations where antibiotics are relatively rare. If we are to resist infections from these organisms in the future, we will need to understand how to avoid the bacteria as well as bringing in new treatments.

A frightening disease which has attracted many headlines in recent years kills quickly, with patients' tissues dissolving into a semi-liquid slush. The illness is known as necrotizing fasciitis, and it is of such recent popularity that it is absent from many reference works. The bacteria responsible are a group of streptococci. These bacteria are common enough, and are responsible for sore throats and occasional cases of septicaemia. Among the well-known diseases involving streptococci are scarlet fever and rheumatic fever, and a sore throat – the so-called 'strep throat' – is familiar to most of us. Not all species are harmful, indeed *Streptococcus lactis*, *S. thermophilus* and *S. cremoris* are important in the production of yoghourt and cheese. The stringy milk popular in Scandinavia is fermented with streptococci. The bacteria grow in long chains, which is why this particular form of yoghourt has its characteristically stringy texture.

There are two factors that make some streptococci a hazard to life. The first is that they can exist as anaerobic organisms without oxygen. Normally they do breathe oxygen, but they can change their metabolism and manage without it. The second is that some strains contain genes that code for the production of highly dangerous toxins. This is a lethal combination: it allows the organisms to grow in tissues which are quickly killed and broken down by the toxins and enzymes that they produce. From this focus of infection, the toxins quickly spread, so a spreading zone of tissue liquefaction soon appears. This would seem to be a case when antibiotics could be used but, even when these bacteria are sensitive to antibiotics, there is no way to get the antibiotics to the site of the infection. Antibiotics are spread through the blood vessels, but the blood vessels around the infection have been broken down and destroyed by the streptococci. The toxins spread further, and the rapid spread of liquefaction moving ahead of the infective organisms makes them impossible to eradicate. The only certain remedy is for the surgeon to cut away the damaged tissues as rapidly as possible and destroy the focus of infection. Necrotizing fasciitis remains a rare condition, and cases are likely to become news whenever they do occur. However, a focus of infection surrounded by a breakdown of tissues requires urgent attention, because death can occur within a day or two unless the process is arrested.

The new diseases we face are mainly caused by familiar types of microorganisms, whether they are recently discovered protists (such as *Acanthamoeba*) or newly recognized bacteria (such as *Campylobacter* and *E. coli* O157). The mystifying Kawasaki disease looks as though it has a bacterial cause, although nobody can yet say what it is. And then we have peptic ulcers and bowel cancer, which now seem to have a newly discovered bacterium as the cause, and coronary heart disease, which may also be caused by an infection rather than stress or diet. In some cases we have yet to discover the cause, in others we are triumphantly claiming to know the reason behind an outbreak: the most frequently cited problem is the overuse of antibiotics. There can be no doubt that we do over-use these compounds, and in a manner that is wanton and frequently related to boosting profit margins.

The greatest world-wide panic was triggered by the discovery, not of a new bacterium, but of a novel virus. Viruses are smaller and simpler than bacteria, and consist of little more than a small package of genes. As a result of this, they cannot reproduce themselves; instead, they take over the genes of a host cell and compel the cell to make more virus. They are the ultimate parasite. There are no antibiotics that can cure a virus disease, because antibiotics interfere with the functioning and reproduction of microbial cells – and viruses are not cells. Smallpox, chicken-pox and mumps are all caused by viruses, and our main line of attack against them remains immunization. There is still no specific drug treatment that will rid the patient's body of the virus. The reason doctors often prescribe an antibiotic (like streptomycin) when a patient is suffering from a virus (like 'flu) is not to rid the patient of the virus, but in the hope of preventing an opportunist bacterium from taking hold of a weakened body and causing a concurrent infection of its own.

The incurability of virus diseases suddenly exploded upon a complacent world in 1983, when a new virus was discovered by a team of virologists at the Institut Pasteur in Paris. It caused a disease that the French workers called SIDA. It is better known to us by the acronym that reveals the order of terms used in English: AIDS (acquired immune deficiency syndrome). The existence of a strange new disease first emerged in New York in 1978 and Los Angeles 1979, when reports were sent to the Infectious Diseases Control Center in Atlanta, Georgia. The New York authorities had noted a sudden cluster of cases of Kaposi's sarcoma (a rare cancer) in homosexual men. In Los Angeles the homosexual community was afflicted by a different complication, infection with an organism known as *Pneumocystis*. The Atlanta scientists realized that the conditions might be related. The sarcoma and *Pneumocystis* are normally controlled by the natural immune system in healthy persons, and it began to appear that both groups of patients were suffering from impaired immunity. By 1982, cases in haemophiliacs began to appear, indicating that contaminated blood might be the route of transmission. Evidence to substantiate this view came from Florida, where heterosexual drug addicts were going down with the disease. Early in 1983, a retrovirus was isolated in Paris by Luc Montagnier, and later that year a virus was also described in America. The American virus, when thoroughly investigated, turned out to be identical with the one already discovered in Paris. It has become known as human immunodeficiency virus or HIV. In 1986 a second genetic variant was found in West African AIDS patients, and HIV-2 has also since been extensively studied.

The viruses belong to the lentivirus group which cause slowly developing diseases. The 'slow viruses' as a group were first recognized by an Icelandic veterinarian, Björn Sigurdsson. In the 1950s, a new disease originating from imported sheep was infecting native flocks. The main symptoms were a wasting paralysis of the hindquarters (*visna* in Icelandic) and breathlessness (*maedi*). In 1962, Sigurdsson recognized that the new syndrome *visna–maedi* was caused by a new type of virus. Since his discovery, we have discovered feline immunodeficiency virus and equine infectious anaemia, both lentiviruses, elsewhere in the animal world. HIV was the first to be found in humans, and our early understanding owed much to the pioneering discoveries of that distinguished Icelandic investigator.

Although the French research teams correctly identified the virus, the American scientists originally concluded that these viruses were related to a virus discovered in Japan in 1981. This was found in blood samples from leukaemia patients. It turned out to be a virus that could implant itself into the host's genes retrospectively, which gave rise to the name retrovirus for the group. This retrovirus was discovered to infect the T4 lymphocytes, a type of white blood cell that mediates immunity in healthy people. This human T-cell leukaemia virus was named HTLV-I. By 1984, the American team headed by Professor Gallo had identified a further type which they named HTLV-III; but the French workers at the Institut Pasteur proved that the so-called HTLV-III virus was actually their already discovered HIV. The Americans were pipped at the post throughout these stages of research, although they do not care to admit it. The first successful test for HIV was the ELISA (enzyme immunosorbent assay) test, which was also devised by the French team in 1984.

HIV spreads from one person to another when bodily fluids from an infected person transfer the virus to someone else. The use of shared needles by drug addicts is an obvious route for infection, and so is the administration of blood products to haemophiliacs, if the HIV has not been inactivated by heat treatment in the course of production. Anal or traumatic sexual intercourse is also liable to cause bleeding, which is why the disease became associated with the homosexual community and rape victims. I published the first chapter on AIDS to appear in a book intended for general readership in 1985, and the conclusions in that chapter are interesting to read in retrospect. I concluded that risks to the medical staff caring for patients were virtually non-existent, and that the spread of the virus among conventional heterosexual relationships was also relatively unlikely. There has been much talk of an increase resulting from heterosexual couplings, but these are more often found in African countries where other sexually transmitted infections are spread at the same time. In cases such as these, the virus can be carried into the bloodstream by an invading bacterium. At the time, there was a truly horrific campaign to warn the public about the dangers of AIDS. The British television commercials were possibly the worst, with apocalyptic visions of granite tombstones collapsing on the hapless viewer. Millions of pounds were wasted on this misguided effort. At the time I was approached by more than one contentedly married couple who, although monogamous and not in an at-risk group, were frightened into using condoms 'just to be safe from AIDS'. In the face of such campaigning, I still felt justified in downplaying the probable extent of this tragic new epidemic. AIDS is an appalling affliction, and I have rarely felt greater sadness than attending support events in California where the shrinking bodies of victims are crying out for help. In some nations where promiscuity has increased over the years, levels of AIDS have reached appalling levels. However, the epidemic has not decimated global communities as was once threatened, and for that we can afford to be relieved.

As I concluded in 1985, the virus of HIV is not highly infectious. Without direct contact of bodily fluids it is not infectious from person to person. Thus, it is not spread by contagion or by inhalation. In this sense we are fortunate. Viruses are great opportunists, and I hate to imagine the consequences if such a dreadful disease were spread as easily as influenza. Lentiviruses are unlikely to acquire that characteristic, because they are not in the same family of viruses as those normally spread by coughs and sneezes, and we should fervently hope that these genes are never acquired by the group. If ever an easily transmissible characteristic is acquired by a lethal virus, this could truly spell the end of entire human civilizations. As we have seen, the end of such great civilizations as the Aztec, Mixtec and Inca peoples was heralded by the highly infectious smallpox brought by European explorers. The invading Europeans were blessed with immunity to the virus, because of lengthy exposure to it in mild as well as severe forms; but to the refined and civilized peoples of Middle America the virus spelled the end of an entire way of life.

The HIV infection concentrates on the T4 lymphocytes, the amoeboid white cells that are key players in our immune responses to disease. The T4 lymphocytes are known as 'helper cells', and they normally hang around in the lymph glands, the blood marrow, spleen and thymus. As the virus takes over, the T4 cells lose their ability to help the immune system, and start to produce waves of new virus which spread further through the body. We are dealing here with a slow virus, remember, and the symptoms take a long time to appear. The disease itself, acquired immune deficiency syndrome or AIDS, rarely appears within six months of infection with HIV and sometimes it takes years to appear. Some patients have been infected for a decade and still remain remarkably fit. The first symptoms are vague and of a kind experienced by most people at one time or another. They include listlessness and tiredness, diarrhoea and a fluctuating fever. The patient's body weight may fluctuate without apparent reason, and appetite may be poor. The lymph glands may be slightly swollen in the groin, armpit or neck. These symptoms may die down before they begin to give rise to real concern, so patients may carry the virus without even realizing it. As the disease progresses, illnesses normally controlled by the T4 cells begin to appear. Kaposi's sarcoma is a form of cancer that appears as purplish growths in the skin, and its occurrence is regarded as a probable indicator of AIDS. *Pneumocystis carinii* infections, which are also associated with AIDS, take hold typically in the lungs. These organisms are tiny protists and their presence is normally detected by a healthy body so that the immune system can eliminate the germs. The AIDS victim lacks this effective response, which is why the disease takes hold.

Research now suggests that HIV may have been around for far longer than we thought. Cases were thought to have occurred in Zaïre in 1976, and tests of serum samples confirmed that the virus had indeed existed in 0.25 per cent of young women in Kinshasa since 1970, without being previously detected. It has been claimed that the ancestral virus was common in colonies of monkeys around Lake Victoria. Since then, serum samples preserved in hospital collections from long-dead patients have been shown to contain evidence of HIV. It has certainly been present at a very low level in Europeans since the 1950s, but it was the expansion of drug addiction and homosexual promiscuity in the 1980s that provided the conditions for the virus to emerge as the cause of a major epidemic.

The genes of HIV show considerable variation. The virus does not have any means of correcting transcription errors when it moves from the independent state to becoming incorporated into the host's genome. As a consequence, the retrospective copying introduces an altered gene for every 1,000 genes transcribed. This may be by design, rather than by accident, because it means that the virus maintains a mixed community of genetically diverse strains. Faced with our vigilant immune system, the virus benefits from a change of genes which makes it hard to eliminate. The reason I feel that the failure to correct genes may be by design is because some of the genes (those in the centre of the virus) remain remarkably unchanged. The greatest variation is found in the proteins that form the envelope around the virus, and it is this external coating that the host's body tries to recognize. By changing its appearance, HIV contrives to remain beyond the reach of the host's immunity. This virus has a limited range of hosts. Since 1982, it has been shown to be transmissible to chimpanzees, but they did not develop AIDS. Treatment remains imponderable. The drug AZT has been most widely used, and many claims have been made for its anti-virus proclivities, but they remain unproved and the drug is normally administered in desperation and for want of anything better. Although the way the virus changes makes the development of a vaccine seem unlikely, there remain hopes that this may yet allow us to conquer this scourge.

Other new virus diseases are marked out, not by a slow and insidious development, but by drama and high fatality. Most serious of all is Ebola virus, which has a mortality rate approaching 90 per cent. The first epidemic took place across Zaïre in 1976. The infection, which seems to have an incubation period of about 4–16 days, begins with dramatic suddenness. The patient develops a raging fever interspersed with chills, with anorexia and aching limbs. This is quickly followed by vomiting, a sore throat, abdominal pain and diarrhoea, and the patients become apathetic, weak and dehydrated. They seem disorientated and desperately ill. Within days the virus attacks the blood vessels and the tissues begin to break down. The patients begin to bleed from the lungs and intestine. Blood runs from the gums and bursts from the skin. The onset of this bleeding phase heralds death, which results from severe shock and fluid loss as the body tissues collapse. There were over 800 victims of the first great epidemic which spread across Zaïre and Sudan. In Zaïre, 88 per cent of the victims died, whereas the mortality rate in Sudan was 53 per cent. In the following year, a single case was diagnosed in Zaïre, and there was a further small Sudanese outbreak on the site of the original epidemic in 1979. The virus has not been found in monkeys, although a related virus was found in cynomolgus monkeys originating in the Philippines. The tragedy gave inspiration to a movie, which simulated horrifying images of the symptoms.

Green monkey disease is caused by Marburg virus, named after the

German town in which it was observed. There were two outbreaks in 1967, one in Marburg and the other in Belgrade, Yugoslavia. The virus originated in vervet monkeys imported from Uganda for experimental purposes. There were 25 cases of infection from the monkeys, of which seven ended in death, and a further six victims who contracted the infection from the first wave of infection. Since then there have been occasional cases scattered across Kenya, Zimbabwe and South Africa. Ebola and Marburg viruses have similar genes, and there is a third member of the group, Reston virus. This is also infectious to humans, but does not produce a severe disease. There have been reports of other types of the same group isolated from monkeys, although there have been no further epidemics among the human population. The symptoms of Marbug virus infection are like those of its relative, Ebola virus. The disease progresses with bleeding from tissues throughout the body, including the liver, the spleen and the kidneys. Vaccines have been tried, but their use does not seem to confer immunity on animals later challenged with the virus. These are highly dangerous diseases, and vigilance will be needed to control future outbreaks.

We have documented virus infections from Africa over the years since World War II. Argentine haemorrhagic fever (AHF), caused by the Junin virus, was identified in 1957 as a potentially lethal disease with a peak incidence in the month of May. In 1959, a similar disease caused by Machupo virus was recognized in Bolivia, and an outbreak of Bolivian haemorrhagic fever (BHF) in 1963–64 caused several hundred cases. There is now known to be another strain, Guanarito virus, identified in Venezuela in 1991, which the perceptive mind will already have recognized to cause VHF. All have natural reservoirs in rodents which live close to the villages. Lassa fever belongs to the same group, and was first identified in 1969. It is endemic to West Africa, from Senegal to Cameroon. I have visited hospital establishments in West Africa where patients with such diseases are recovering, and it is haunting to realize what privations these epidemics can cause. There are resonances of earlier eras in Europe when diseases such as the 'Black Death plague' (still found in the USA) and smallpox still held dominion over the human population. Many nations daily face conditions that we feel proud to have conquered in our own societies. These haemorrhagic fevers, including Lassa fever and AHF, involve bleeding from the mouth and internal organs and massive damage to the liver. They all have a mortality rate of about 16 per cent and, although most patients recover, the process of recuperation is long and distressing.

A major group of viruses causing the breakdown of internal tissues are the hantaviruses, the cause of Korean haemorrhagic fever (KHF). There is little doubt that the disease has existed for centuries, but it was first recorded in Western medical literature in 1951, when there was an outbreak among UN military personnel stationed in Korea. Over 3,000 persons were struck down during the outbreak, with a mortality rate approaching 15 per cent. The cause was identified by Ho Wang Lee in 1978, and named Hantaan virus. It was first cultured in 1981, and international research has revealed that it causes widespread disease throughout the Far East, including Korea, China and eastern Russia. The disease has an incubation period of one to two weeks, before the sudden onset of a high fever accompanied by severe pain in the muscles and eyes, followed by widespread bleeding and the swelling of tissues through oedema. One-third of the fatal cases die during this phase from the effects of severe shock. Kidney malfunction follows and heart failure may become apparent. After a period of severe ill-health the kidneys may suddenly resume normal function, and the patient's probable recovery is marked by the passing of large amounts of urine as the excess tissue fluid is eliminated from the body. Recuperation lasts several months.

The virus is found in rats and mice, and it seems that the disease is spread through contact with the dust from infected rodent droppings. There are some vaccines now under test in the Far East, although sound evidence of their value has yet to be published. Of course, you do not need to find the virus in order to detect where it has been. We can detect where a virus has spread by looking for antibodies to the virus in blood samples. Interestingly, although there have never been any outbreaks documented in the USA, you can find blood samples from people and from rats which test positive for the antibodies to these viruses.

New viruses are regularly recognized as research progresses. In many cases, we had evidence that there were undetected causes of disease, long before the viruses themselves were discovered. In the late 1940s, trials took place in America to demonstrate that diarrhoea could be caused by infections that were nothing to do with microbes. Experimenters filtered samples of faeces from patients with intestinal disease, but filtered the fluids so that every particle (including every microbe cell) was removed. When samples of each were spread onto the standard culture media for bacteria, there was no sign of growth: the filtration had removed them all. None the less, when these were administered to volunteers, symptoms of the gastroenteritis quickly appeared. In this way scientists speculated on the existence of viruses causing diarrhoea, even though nobody knew what they might be. The disease agents were given names from the hospitals where the experiments were conducted, and so we had the Marcy factor (after the Marcy State Hospital at Utica, New York) and the Niigata factor (Niigata Prefecture, Japan). Even a generation ago, nobody had identified a virus as the cause of gastroenteritis; indeed the very first paper describing a virus in a case of diarrhoea did not appear until 1972. This resulted from an outbreak of 'winter vomiting disease' which struck down half the teachers and pupils at an elementary school in Norwalk, Ohio. Volunteers were administered filtrates from these patients, which showed that a bacteria-free extract could transmit the disease. This time, the electron microscope was used to examine concentrates from these samples and tiny virus particles were seen for the first time. The technique soon revealed a host of new viruses, including hepatitis A. A year later the rotaviruses were spotted, and they are now known to be causing a world-wide epidemic of viral gastroenteritis. What we call 'gastric 'flu' has since become well known, although to delicate or malnourished children these diseases can be life threatening.

Five million infants die every year from diseases that cause diarrhoea, and we are just beginning to recognize that viruses lie behind many of them. For example, one-fifth of all those deaths are caused by the rotaviruses. That's a million deaths each year caused by a germ that we have only recognized in recent years. Rotaviruses are found widely in animal and human populations. In Europe, the outbreaks are usually concentrated in the cooler months, whereas in the USA there is a peak of infection that begins in the south-west every November, and moves up and across the States until it reaches New England in March. In the tropics there is no seasonal peak, and cases occur all year round. Infection with these viruses in children below the age of two months, or older than two years, seems to produce few symptoms. It is between these age groups that the virus produces its worst effects. Susceptible adults sometimes contract the virus too, and in them it can produce gastroenteritis. Among the children who become ill, most have diarrhoea, high temperature and vomiting. One in a hundred produces blood-stained faeces, and in the Western World a third of the children actually end up in hospital. Vaccines have been tried, but they do not always work. There has been much emphasis on the existence of antibodies against rotavirus in breast milk, but research has not reliably confirmed that breast-fed babies are really less likely to become ill. The standard treatment for these illnesses is rehydration with a weak sugar solution in saline, but this is not the cure-all that it is sometimes said to be. Weak children given too much rehydration therapy can build up such levels of sugar and salt in the intestines that water is removed from the body even faster than before. There have been cases where paediatricians have had to cope with a child having seizures as an additional complication – the result of too much rehydration fluid administered by anxious parents. If we could develop a reliable vaccine (the best trials have shown no more than 70 per cent effectiveness), then we could greatly reduce the incidence of childhood gastroenteritis. Here we have a chance to curtail a tremendous burden of world-wide suffering, much of it in developing nations. Our identification of new viruses offers some promising challenges for the future, with world-wide life saving as the clear result.

Finally, let us look in detail at a new category of diseases which we are struggling to understand. From these new facts we can learn lessons that may be crucially important as the new millennium unravels. Humankind is now facing a new disease that does not fit into any of the conventional categories: bovine spongiform encephalopathy (BSE). There have been many theories to explain this epidemic among British cattle, and we still need to know more. It is a fatal affliction of cattle from which recovery is unknown. Hundreds of thousands of cattle have been slaughtered in a belated attempt to conquer the disease, which offers many lessons in the way epidemics of novel diseases should be handled in the future. We have so far encountered new diseases caused by a range of microorganisms, but BSE does not appear to fit into any of the categories. According to current theories, the causative agent is not a bacterium, it is not even a virus, but takes the form of a highly resilient infectious protein dubbed a prion. It contains no genetic material at all, and is unharmed by conventional disinfectants and by heat sterilization. The lessons here are important: genes cannot hold the ultimate secret of replication, for here is an agent that lacks genes of any sort. It is also clear that BSE can pass to human victims. The disease has attracted world-wide interest, partly because of the unique factors which make us fear it so much:

- First is its nature. This is not an illness that simply gets better. In this dreadful disease, the mind collapses. Holes appear scattered through the brain tissue. There is something peculiarly threatening about an attack on the brain.

- Second, there is the outcome: every case is fatal. There have never been any recoveries. Death from the illness is always certain.
- Third is its insidiousness. This is a disease that may be passed on through doing what most people have done for thousands of years – eating beef. It doesn't come from bad hygiene or carelessness, like food poisoning.
- Fourth is the mysterious origin. This is a fascinating condition. There is no microbe as the cause. There isn't a virus for 'mad cow' disease. The public has been watching the unravelling of the structure of DNA reported from laboratories around the world. Every known form of infection has always had DNA (or RNA, its 'messenger') at its heart. But these new diseases do not even seem to have that. This is a completely novel form of infective agent.
- Then there is the role of the advice people have been given. In specialist subjects the public like to take the best advice available. The Government have insisted for ten years that there is no risk to humans. We have to heed the advice of scientists, they said. In science we knew that there was no evidence of transmission to people. There still isn't any evidence that BSE from cows can cause CJD (Creutzfeldt– Jakob disease) in humans – but many words of warning were given.
- The sixth major dimension of this historic episode is the sense of fear in the public. People do not know where to go, where to turn, whom to ask. Public information has been badly handled. The information that has been released is misleading, and the facts have been concealed.

The unravelling of this new epidemic began at Easter 1985 when a veterinary surgeon, Colin Whitaker, was called to a farm in the 'garden of England', Kent, to see a sick cow. Normally a docile and compliant animal, she was becoming hyperactive and aggressive towards the farmer. The disease resembled 'the staggers', which has been known for centuries. It results from a lack of magnesium in the diet, and it can be cured by treatment with supplements. Mr Whitaker soon realized that he was witnessing something very different: it looked like scrapie – a disease of sheep suddenly infecting a cow. The case was brought to the attention of the Central Veterinary Laboratory in Weybridge, Surrey. One of their pathologists, Dr Gerald Wells, concluded that this was a previously unrecorded condition – bovine spongiform encephalopathy, or BSE for short.

The most famous victim of BSE was Daisy. She was a black and white Friesian suckler cow aged six years. The Friesians are the most common milk cows in Britain. Daisy is the unfortunate beast who falls about so dramatically in the early video of 'mad cow' disease which regularly re-plays on television. Her uncoordinated gait and poor bodily condition were reflected in an agitated and nervous manner. She frequently tried to rub her head, either with a front hoof, or against a wall, and soon lost the ability to walk. The videotape provides a graphic illustration of the terrible consequences of contracting spongy brain disease. The youngest animal ever to contract BSE was aged 20 months, the oldest 18 years. However, we believe that it has a normal incubation period that seems to range between two and a half and ten years.

Could we learn lessons from the existing knowledge of scrapie in sheep? The first cases of scrapie in the English-speaking world probably occurred in Merino sheep imported from Spain in the fifteenth century. Since then it has been given a variety of names, 'animal rickets', for example. It has even been dubbed 'distemper', which is a disease of domestic animals with no relationship to scrapie. In 1759 it was named *Der Traberkrankheit*, the 'trotting disease', by an investigator named Leopold in Germany who recorded this classical description:

Some sheep also suffer from scrapie, which can be identified by the fact that the animals lie down, bite at their feet and legs, rub their back against posts, fail to thrive, stop feeding, and finally become lame... This illness is incurable. The best solution, therefore, is for the shepherd to dispose of a suffering sheep quickly, slaughtering it away from the manorial lands. A shepherd must isolate such an animal away from the healthy stock immediately, for it is infectious and can cause serious harm in the flock.

Scrapie has since been reported right across Europe, in the Netherlands and Italy, Switzerland and Spain for example. It is present in Japan, and has spread in countries including Iceland, Norway and Cyprus, after being imported in infected animals. It is still absent from some countries including Australia, New Zealand, Argentina and Uruguay. The cause of scrapie has been variously identified over the centuries. Although scientists don't often realize it, changing theories usually mirror the social preoccupations of the time. In the middle of the nineteenth century everyone was concerned by issues of morals and sexuality, so in 1848 it was suggested that scrapie was a mental degeneration caused by the sexual excess of the animals. Later on, electricity became fashionable and it was claimed that scrapie resulted from lightning. By the turn of the century, the germ theory of disease was popular and the 'germ of scrapie' was claimed to be *Sarcocystis*, a microscopic parasite of animals.

In the 1940s a research scientist lost his reputation because of his theories. Dr D.R. Wilson was fascinated by the disease and came close to recognizing its unusual nature. He proved that the causative agent was smaller than microbes, and was strangely resistant to heat and disinfectants. Others refused to believe his results, and he has only posthumously been recognized as an honest observer. After Björn Sigurdsson came up with the new idea of 'slow viruses', it was believed that one of these might cause scrapie. The true cause took over two centuries to emerge. Scrapie is one of the degenerative spongiform diseases – the first to be identified, but certainly not the last. We now know of others. Chronic wasting disease (CWD) was discovered in 1980 in the elk and deer populations of North America. In mink we now recognize transmissible mink encephalopathy (TME), which makes them excitable and over-reactive. We already know of a similar disease of ostriches, first reported in 1986. The most recent new disease to be reported is feline spongiform encephalopathy (FSE) which was discovered in Bristol in a cat named Max. Within four years over 50 cases had been identified, and the disease was then found in zoo animals including pumas, cheetahs and ocelots. This disease seems to have been triggered by the eating of pet food containing BSE-infected meat. It does not transmit disease to all cats, however; lions and tigers seem to be immune to the disease, although they have eaten plenty of potentially infected carcasses.

So where did BSE originate? Is it possible that it may have existed for centuries, and had simply been missed? There was an interesting candidate case in France as early as 1829, where a naturalist named Berger published an account of a disease that is very like BSE. There have been reports of 'scrapie' in oxen dating from before World War I in the English Lake District. Several reports have circulated of cases of scrapie in cattle seen in Cornwall years ago, and of cows with abnormal behaviour in southern England. We have been taught that the disease arose from cattle feed contaminated with scrapie, but I do not want us to lose sight of the possibility that it arose from a cattle carcass infected with age-old BSE, which was spread to other cattle by being rendered down and included in the mix.

Science has known for years of several human spongiform diseases. The

---

Table of spongiform diseases

Human:	Creutzfeldt–Jakob disease
	Gerstmann–Sträußler–Scheinker syndrome
	Kuru
Animal:	Scrapie – in sheep, goats and moufflon
	Bovine spongiform encephalopathy
	Transmissible mink encephalopathy
	Chronic wasting disease in deer and elk
	Feline spongiform encephalopathy (cats, cheetahs, etc.)
	Spongiform encephalopathy in zoo antelopes (kudu, oryx, etc.)

From BSE: *The Facts*, by Brian J Ford. London: Transworld, 1996.

---

best known is Creutzfeldt–Jakob disease (CJD), named in 1922 after two German medical researchers: Dr Creutzfeldt (who recorded the first case in 1920) and Professor Jakob, who made a follow-up study of further patients with the disease. Apart from three clusters of CJD that have been reported in Libya, Slovakia and Chile, it is known that the incidence in the human population is always one or two per million. This is the same in countries like England (where scrapie occurs) and Australia (where it doesn't), suggesting that CJD is not contracted from scrapie infections even though the diseases are very similar. Have we perhaps underestimated the incidence of CJD in the past? Collections of microscope slides from the past can help answer this basic question. A survey of existing brain sections could reveal whether there were such patients. Dr Gareth Roberts of the pharmaceutical company SmithKline Beecham has looked at 8,000 brain sections – mostly from the Corsellis collection at Runwell Hospital, Essex – and narrowed them down to 1,100 from people who had died of dementia. He found microscopic evidence of CJD in just 19 cases, 11 of which had been correctly diagnosed during life. Eight had been missed. It may be that the 'new' form of human spongiform encephalopathy has been around for years. There have been unexplained cases in the past, which psychiatrists could not identify.

We call this disease 'new-variant Creutzfeldt–Jakob disease', usually printed as nvCJD to make it look sufficiently cryptic. I do not believe that this can be justified, because, to be a subset of an existing disease, the 'new variant' must be linked to whatever disease it most closely resembles, otherwise you might as well describe a chihuahua as a new variant of kitten. There is a disease that it more closely resembles: kuru, the smiling sickness of the Fore people of Papua New Guinea. They have long possessed herbal remedies from the forests with which to tackle diseases, and know much about the uses of plant medicines. One disease, though, could not be treated by the traditional remedies of the tropical rain forest. The symptoms were similar to malaria. Patients tremble and have difficulty holding things or walking. After a few months they could not even stand, and they lay where they were with eyes staring, sometimes grinning insanely yet unable to communicate, until they died. It was a tragic and terrible end. The local population called this 'laughing death'. The explorers were told that these poor people were shivering: *kuru* in the local language, so kuru is how the disease is known to Western scientists. It decimated villages. Victims could be seen lolling about, crawling across the floor, or shivering and smiling strangely in their beds.

The first scientific accounts of kuru were published by Professor Carleton Gajdusek in 1959 and won him the Nobel Prize for Medicine and Physiology in 1976. Gajdusek liked the traditional way of life in Papua New Guinea, living in homosexual communities with boys, and was later arrested on charges of paedophilia when he transferred the same habits to America. During his sojourn he recorded that the tribes celebrated the death of elderly relatives by opening the skull of these newly deceased persons and devouring the brain, or smearing it on their bodies. There were over 2,500 victims of kuru in the Eastern Highlands of Papua New Guinea between 1957 and 1975: one person in 100 died each year. In 1960, cannibalism was outlawed, so the incidence of kuru began to subside. One or two cases are still found, but no patient aged under 22 years had died since 1973, and no case of kuru has been recorded in anyone born after 1959. There are rumours that the old practices are beginning to reappear (this is the era of cultural self-expression), so it would be too much to claim that kuru was entirely extinct, even at the turn of the new millennium.

Although Western cannibalism is rare, cases of CJD have been transmitted from person to person through the transfer of tissue and tissue extracts. These include:

- Corneal grafts in ophthalmic hospitals
- Electrodes experimentally implanted in the brain
- Grafts (injections) of brain cells in degenerative diseases
- Instruments used in surgical operations on the central nervous system

- Human gonadotrophin
- Human growth hormone.

The spread of the disease through the use of properly sterilized surgical instruments, or the administration of hormones extracted from diseased pituitary glands, has a haunting poignancy. CJD is not the only spongiform encephalopathy found in the human population. A rarer disease is Gerstmann–Sträußler–Scheinker syndrome, first described by Dr Gerstmann in 1928 in Vienna. Eight years later, with his colleagues Messrs Sträußler and Scheinker, he published a paper on the syndrome. The disease can take over 10 years to develop. It is an extremely rare disease, with one case in every 50 million of the population. Even less well known is fatal familial insomnia (FFI), which seems to be the result of a rare genetic mutation.

The new disease from British beef was first suspected in 18-year-old Victoria Rimmer of Connagh's Quay in north Wales. She became ill in 1994, and 30-year-old Maurice Callaghan died of spongy brain disease in Belfast during the following year. His grave was reportedly dug nine feet deep, rather than six, and it is said that the grave diggers wore protective clothing and rubber gloves while dealing with his interment. Jean Wake of Sunderland died at the age of 38 with her children and mother at her bedside. Her mother, Mrs Nora Greenhalgh, was convinced that infected beef had led to the death of her daughter. She wrote to No. 10 Downing Street, to draw the attention of the Prime Minister to her belief. The reply came from Ms Rachel Roberts, private secretary to the Prime Minister. It was quoted in *The Times*: 'I should make it clear that humans do not get "mad cow disease", although there are similar diseases which naturally occur in humans and have been known about for very many years,' she wrote. 'I must reassure you that there is no evidence to suggest that eating meat causes this sort of illness in people.' Stephen Churchill, a promising student, died in May 1995 aged 19, after a year-long illness marked by dizziness and depression and, in Manchester, 29-year-old Michelle Brown died in November that year shortly after giving birth to her son. In February 1996, a student from Chester-le-Street in County Durham, Peter Hall, died of the same condition just before his twenty-first birthday. Within a year there were over ten cases of this new form of spongy brain disease known to the authorities. There was a spurt of five new cases between March and August 1997, taking the total to 21; by Christmas that year there were 23 in total. On

26 March 1998 (two years to the day since the Government finally conceded that the new disease really did exist) the twenty-fourth case was confirmed. During the time that the disease was being hidden from the public, the scientists working in the field kept insisting that there was little or no evidence of risk. Ever since disclosure, the same people have been warning of a forthcoming epidemic of 'biblical proportions'. This I doubt: 24 cases over four years makes the new disease one of the rarest in existence. Whereas the official spokesmen have their future employment to consider, I am constrained only by the simple laws of biology – and such a small incidence, without any upward turn in the graph after such a long period, does not cause me undue concern. We face far greater risks every day of our lives, by doing such hazardous things as walking up and down stairs. Do you drive a car? You have a chance 1,000 times greater of being seriously injured or killed than those tragic young people had of dying from eating beef. Do you travel by train? There is ten times as much chance that you'll be killed in a railway accident within a year.

I remain convinced that calling the new disease a variant of CJD cannot be justified. Creutzfeldt–Jakob disease is a different syndrome with characteristic signs and symptoms. In CJD, forgetfulness is an early sign; in this new disease anxiety and uncharacteristic depression first appear. Victims of classic CJD typically succumb within six months, whereas in the new cases they survived for more than a year. The microscopic appearance of the new cases was distinct from what is seen in CJD, and has more in common with brain sections from victims of kuru. The average age for the onset of CJD is over 60; in the new cases the average age was below 30.

There is now much evidence to suggest that prions could be the cause of the spongy brain. One early definition said that it was derived from protease-resistant protein. A more recent description, published in April 1996, said that it came from a proteinaceous infectious particle. I doubt whether the latter etymology makes any sense at all. Every germ is an infectious protein particle, and the point about the prion is that it resists the normal enzymes (proteases) that do attack proteins. That's why prions are hard to destroy. Prions are produced in normal brain cells. The code in the gene that produces prion protein (the PrP gene) is translated into messenger RNA (mRNA) to assemble the molecules and generate the prions. The theory suggests that some forms of prion protein fail to fold properly, but grow at unaccustomed angles, resulting in damage to the brain cells. There are 23 pairs of chromosomes in each human cell, and the genes that make prions have been located on the short arm of chromosome 20. There are genetic mutations already known in this area. Some of them are associated with human disease. One has already been found in several families with fatal familial insomnia.

What lessons can we learn from the BSE crisis? What should be our aim if we are to avoid facing such terrible risks in the future? The spongy brain diseases have some characteristics in common with other diseases, such as Alzheimer's disease. In some cases of this disease, we find evidence of a substance known as amyloid precursor protein (APP). This seems to be faulty mRNA. Deletions of guanine and adenine seem to be the cause, and we may yet find that our ability to sequence genes will help throw light on terrible diseases such as these. Alzheimer's disease is very common. In Britain alone there is a death from this condition every five minutes. It is like a plague on modern society and threatens us all.

The urgent research into BSE and the spongiform diseases of humans may teach us lessons that could apply to these other conditions. The air of secrecy and incompetent planning has certainly some lessons to teach us if we are to avoid further catastrophes in the future. The targets for which I believe we should aim have been listed like this:

- To banish spongiform diseases. We must tolerate no more massaging of facts. Slaughtering animals before they develop symptoms must never again be considered. BSE must be eradicated, with scrapie next on the list.
- To restore openness in science and medicine. Never before has the Government interfered so dangerously with the free discussion of results. The hazards of the past – like tobacco, and the problems over saturated fats – have always been fully aired. People make up their own minds.
- To facilitate access to knowledge. Scientists must express their views openly and directly. The truth about BSE was concealed. In some parts of the world it is unlawful to mention to a sexual partner that an individual has AIDS. If we need laws in such areas, they should ensure that facts are known, and not concealed. Lives hang in the balance through this growing form of secrecy.
- To understand that the creation of knowledge, wisdom and insight is the guiding principle of scientific enquiry. Creation of wealth, the

guiding light of modern science funding, is irrelevant. Greed is anathema to science.

- To make that great leap, we need a paradigm shift away from the philosophy of 'cover up and carry on as normal'. If we have public health and well-being to consider, epidemic threats must be eradicated at the outset. The extirpation of disease must apply to BSE; it should then apply to scrapie; it might usefully be extended to *Salmonella* in chickens.
- Feed must be germ free. We would not tolerate infected food at home, simply because too many people in the family go sick. We require perfectly wholesome food at all times. So do farm animals. The cost of sterilization is minimal and germ-free food is easy to obtain.
- No longer must we tolerate the burden of the degenerative diseases. There is a vast spectrum of conditions inflicting suffering on a billion people. Efforts should go into drawing them together in a network of understanding.
- Pure research requires pure funding. Privatization may well work for meteorology or high-energy physics where there is a product in view. But visionary science at the leading edge of thought demands continued funding free from commercial pressures. Anyway, pure science in biology is not expensive.

There is much that we must learn from this disastrous outbreak. The British Government has intervened in the open publication of science, censoring news and laundering the facts in order to protect their own priorities. The fact that there were risks was no reason to bring down a cloak of secrecy. We take chances every day. Eating is a risky business and it always has been. You may choke on an olive, contract cancer from celery or coronaries from cake. There are cancer-causing chemicals in cress and kippers, and the risk of death lurks in a can of lager. Living means we die, and our concern must not be allowed to transmute into unreasonable fear. Eat too much and you are threatened with obesity, too little and there could be a risk from anorexia. Human history has been marked by food scares and by fashionable beliefs about diet.

In the end, science owes it to the public to be honest and straightforward about its findings. We are creating new diseases as society develops, and the facts have to be clarified as a matter of top priority. In future we must never allow the truth of science to be subverted by the expediency of governments on a short fuse.

7

The good microbe

PASTEUR DID US ALL A GREAT DISSERVICE. I do not imagine that he did it deliberately, but his germ theory bequeathed to us a fear of microbes. We are left with the impression that a microbe is a germ, and that a germ causes disease. Remember what I said early in the last chapter: *disease germs are maverick organisms*. So they are. The overwhelming majority of microbes are harmless, and there may well be more organisms positively promoting health than there are germs causing disease. The disease germs are pathogens (from the Greek *pathos*, meaning disease). In an article for the scientific journal *Nature* in 1975, I introduced a new term for the organisms that make us healthy. To match pathogens, I proposed to call them *salugens* (from the Greek *salus*, meaning health). We must not be seduced by the idea that microbes are bad for us. On the contrary, they are vital for life. Why this strong association between the word 'germ' and disease? The word means a living essence, not an unhealthy principle. The earliest uses of the word all referred to this property of bursting with new life. In English literature, the first time the word appeared in print was in 1644, and every use of the term concerned the germ as the living principle from which new organisms were born. 'Germ cells' are reproductive cells. The 'germ of an idea' is the essential seed from which it springs. 'Germination' is the sprouting of new life. 'Wheat germ' is the embryo within each grain which is the centre of the plant's reproduction. None of these familiar uses of the word relates to anything negative, or harmful; and the earlier uses of the term were closer to reality than the narrower meaning that we attach to it today.

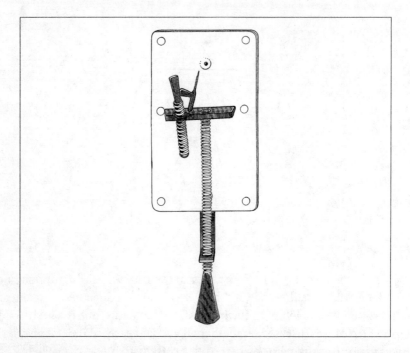

Leeuwenhoek's pioneering microscope. With little microscopes like this (each the size of a rectangular postage stamp), Antony van Leeuwenhoek discovered the world of micro-organisms. After working as a draper and town official, he started microscopy at the age of 40. In spite of his late start, he spent the next 50 years making detailed observations on microscopic life.

When the Dutch draper Antony van Leeuwenhoek first saw microscopic organisms in the specimen of lake water that he collected from Berkelse Mere in late August 1674, he wrote an account which betrays the sense of wonder he experienced. Here is an extract. Just read the excitement and delight at the moment Leeuwenhoek invented a new science – microbiology:

Passing just lately over this lake, at a time when the wind blew pretty hard, and seeing the water as above described, I took up a little of it in a glass phial, and examined this water next day. I found floating therein various earthy particles, and some green streaks, spirally wound serpent-wise, and orderly arranged, after the manner of the copper or tin worms, which distillers use to cool their liquors as they distil over. The whole circumference of each of these streaks was about the thickness of a hair of one's head. Other particles had but the beginning of the aforesaid streak, but all consisted of very small green globules joined together, and there were very many small green globules as well. Among these there were, besides, very many little animalcules, of which some were roundish, while others, a bit larger, consisted of an oval. On these last I saw two little legs near the head, and two little fins at the hindmost end of the body. Others were somewhat longer than an oval, and these were very slow moving and a few in number. These animalcules had different colours, some being whitish and transparent, others with green and very glittering little scales, others again were green in the middle, and white at the front and rear, others again were ash grey. And the motion of most of these animalcules in the water was so swift and so various, upwards, downwards, and round about, that 'twas wonderful to see: and I judge that some of these little creatures were less than a thousand times smaller than the smallest ones I have ever yet seen, on the rind of cheese, in wheaten flour, mould, and the like.

If there is one intriguing fact about Leeuwenhoek's life, it was that he paid so little attention to microbes as pathogens. Many commentators have since criticized him for his narrow vision, and believe he could have launched the germ theory of disease some two centuries before Pasteur. I can find little justification for this view. The idea of infection dates back long before Leeuwenhoek. It occurs in the Bible. In the Book of Leviticus (700 BC) there is an account of leprosy, and the idea that sufferers from disease might be contagious. The Roman encyclopaedist Marcus Terentius Varro (116–27 BC) wrote of the possibility of 'germs' spreading disease around 50 BC, and in 1348 an epidemic in Florence was described as being spread by an unseen agent in contaminated clothing. This account was written by Giovanni Boccaccio (1313–75) in his celebrated *Decameron*. In 1546 Girolamo Fracastoro (1478–1553) published *De contagione et contagiosis morbis* in Venice. This set down the principles of the germ theory. Epidemics, he said, resulted from rapidly reproducing minute agents which could be spread in three ways: by direct contagion, through carriers such as infected clothing and through the air. Athanasius Kircher (1601–80) wrote 44 books, including *Scrutinium physico-medicum contagiosae Luis* in 1658, which referred to contagious effluvia containing microscopic living bodies. He wrote in vague terms, speaking of 'very fine and subtle and very light' germs that could 'insinuate themselves into the innermost fibres of clothes'. Pierre Borel, in 1665, wrote that minute 'worms' were found in the blood of people suffering from fever, and then a second edition of Kircher's work was published in 1659 by Christian Lange (1619–62), which gave the idea of a germ theory of disease a further boost in popular interest. By the time Leeuwenhoek first looked through a microscope, when he was 40, the idea of dangerous germs was of considerable antiquity, and was widely known. Indeed, his observation of spermatozoa in human semen seems to have resulted from the idea that they may have been the causative organism of gonorrhoea.

I believe that Leeuwenhoek was well aware that microbes might cause disease, but had a much more balanced view: the microbes he was examining were beautiful and marvellously interesting, and these were the ones worth studying. If there were disease germs, then they were something different. He did not write about pathogenic microbes, because he never studied any. If this is the right interpretation, then his view of the microbe world was more balanced than the way science has regarded them for over a hundred years. The fact is that the majority of organisms around the world are microbes, and the majority of microbes are valuable members of the global community. These are the organisms that we need to study if we are ever to grasp the way the environment works. Leeuwenhoek was the first to study them, and in the 1600s he pointed the way in which science must look at life in the twenty-first century. Microbes are more rationally viewed as a powerful force for the common good. They occur in more environments than any other form of life; indeed, the populations you carry with you are far larger than you might expect.

You may well have taken a shower a little while ago. You used soap or a gel which promised to rid your body of bacteria. Bacteria make you smell, they say, and they also make you ill. Well, let's take a brief tour around your body and look at the microbes that are teeming all over you, both inside and out. No matter how thoroughly you scrubbed yourself, you are covered with bacteria. Almost all of them are friendly. There are trillions of *Staphylococcus albus*, a harmless relative of the *S. aureus* we encountered earlier (see page 114). Both staphylococci are normally found on healthy human skin, and there are vast numbers of *Sarcina*, another spherical bacterium which grows in neat cubic packets of eight. You will also find countless numbers of the diphtheroid group of bacteria, and masses of yeasts, such as *Pityrosporum ovalis* which is particularly common on your scalp. The yeasts consume the oily fat that you secrete each second of life. These organisms, and many others, form a complex community that resists our efforts at hygiene. The most thorough washing and bathing has remarkably little effect on these innumerable microorganisms, which greatly outnumber the entire human population of the world.

There are billions more diphtheroids around your genitals, and a relative of the tuberculosis germ poetically named *Mycobacterium smegmatis.* Inside your nose is much the same selection, with rather more *Staphylococcus aureus,* which prefers this moist environment. Back towards the throat there are many streptococci, notably *S. salivarus,* and *Neisseria pharyngis.* Its cousin *N. meningitidis,* which causes meningitis, is quite common in the nasal passages, along with organisms that can cause chest disease including *Haemophilus influenzae* (which was believed to cause influenza until the 'flu viruses were discovered) and *Klebsiella pneumoniae,* which can cause lobar pneumonia. The structure of the mouth and nose manages to control most of the organisms that might get into your lungs, and in healthy people the contents of the lungs are remarkably free of microbes. The bronchi are lined with a flickering layer of cells covered with cilia, and these beat incessantly upwards, causing a current of mucus which perpetually rises towards the throat. Any bacteria that are caught on this layer are wafted upwards, to be swallowed with saliva where they are faced with an acid bath in the stomach.

In young girls and postmenopausal women the vaginal environment is slightly alkaline, and a huge variety of skin organisms exists within. There are the usual staphylococci and streptococci, along with yeasts like *Candida* and bacteria of the *E. coli* group. Many of these bacteria produce compounds that control the pathogens; they are truly salugenic organisms, and help to keep us healthy. During a woman's fertile years, the environment of the vagina becomes acidic, rather than alkaline, and the population of microbes changes to suit the different environment. In adult women the predominant bacterium is *Lactobacillus acidophilus,* an organism also found in the production of sauerkraut and silage, pickles and cheese. There are large populations of club-shaped bacteria of the genus *Corynebacterium* (some species of which cause diseases in farm animals as well as humans) and a selection of streptococci, particularly the species that can exist without an oxygen supply. Yeasts such as *Candida* often prefer an acid environment, which is why women sometimes suffer from thrush (called candidiasis in the medical world) if this organism gets out of equilibrium with its neighbours. The mouth receives new microbes with every breath you take. It is rich in nutritious debris, a mix of food particles and dead cells shed at the rate of millions a minute from the mucous membranes, which line the mouth and gums. For a bacterium, this is a marvellous environment – it has endless supplies of food and water, liberal oxygen and perfect warmth for growth. There are fine hair-like bacteria of the genus *Leptothrix* which coat the teeth, and darting in and out are minute spiral corkscrew-like spirochaetes, harmless relatives of *Treponema pallidum* which causes syphilis. There are many yeasts, including *Candida* and *Pityrosporum*, and species of *Neisseria* and *Fusobacterium* which proliferate between the teeth. You may brush your teeth for an hour, or all night long, but you have negligible effect on these organisms. It is just as well, for you rely on them for health. Face it: with its perfect position for collecting microbes, and its liberal supplies of everything a bacterium could need, your mouth ought to become a putrescent and decaying mass of microbes, all clamouring for more. In fact, even if you bite your tongue you will probably be surprised how quickly and painlessly it heals. The worst most people suffer is gingivitis, a chronic infection of the gums resulting from a build-up of tartar around the base of the teeth. Why does the normal mouth remain so healthy? It is not because we keep the microbes out, by dental hygiene, but because the microbes live there in such vast and irrepressible numbers. It is our salugen populations, those teeming billions of protective microbes, that help maintain your health.

Few of these bacteria ever get through to the intestines, for most are killed by the bath of strong hydrochloric acid that lies gurgling in your stomach. This acid, strong enough to burn a hole in the carpet, helps break down food and is also an important means of killing germs. Few of them survive passage through this hostile environment. However, when we enter the intestines the population of organisms changes dramatically, and becomes enormously increased in number. There are more bacteria in a single spoonful of the contents of your lower intestines than the total number of different species of life everywhere in the world.

Where did they come from? The intestines of a newborn baby are completely sterile, and the only organisms that you can find are on the baby's skin and in its mouth, where it is heavily populated with the same microbes found in the mother's vagina. This is an unacknowledged advantage of natural childbirth, where possible, for the inoculation the baby acquires as it passes through the birth canal establishes an already balanced microbial community on its skin. Within a few days, the balance of this microscopic population shifts and becomes more like that of an adult. Bacteria are acquired from handling and from the breath of the parents, and the normal baby soon makes the transition to life in the open air. Meanwhile, bacteria start to proliferate in the intestines. A few *Lactobacillus bifidus* arrive from the mouth, because they are able to resist the hydrochloric acid through which they must pass on the way down, and these bacteria make up most of the microbial population within the intestines of little babies. Most of the other species get in through the anus and start to grow upstream, against the downward movement of digested milk products and shed intestinal cells that the baby is continually processing. It is a remarkable period of colonization, and it is not until the baby is weaned that the community of intestinal microbes is finally established.

The microbes within the human intestines are a varied lot. There are huge populations of *Escherichia coli* in carnivorous and omnivorous mammals, although these bacteria are in the minority in herbivorous species. In guinea-pigs, for instance, lactobacilli make up 80 per cent of the bacteria in the gut. Growing alongside the *E. coli* in the human intestines are different species of *Bacteroides*, lots of different species of cocci and bacilli, sizeable populations of clostridia, particularly *Clostridium perfringens* – previously known as *C. welchii*, after its discoverer William Welch (1850–1934) who was a professor at the Johns Hopkins Medical School in Baltimore, Maryland. This is a relative of the botulinum bacterium, and is familiar as the organism causing gas gangrene. *C. perfringens* is widespread in the soil, and there is no way we could hope to avoid it. Fortunately, it rarely causes any problems and may even contribute to our inner state of health. We do not yet know the full extent of the species that live in our intestines, but there are vast numbers of bacteria of almost every shape and size. Between them they help process our food, provide us with vitamins, and maintain a generally healthy environment in which we can digest our food and absorb the results. The only time that we are likely to be reminded of their importance to us is when we take oral antibiotics. These knock out some of the bacteria, and the community becomes upset. During a period of antibiotic therapy people can develop fungal infections, such as thrush caused by *Candida albicans*. As we have seen, an infection with *Clostridium difficile* can also result from antibiotic therapy. It is the removal of salugens that allows these potential pathogens to come to the fore.

We owe so much to the proper use of antibiotics. Yes, I know it is fashionable to say that we would be better off without them, but anyone who has suffered bacterial meningitis or tuberculosis, or who has had a child suffering earache or tonsillitis, knows how much benefit antibiotics can bring when properly used. We have seen the range of antibiotics on pages 113–15, and need to remind ourselves of one crucial fact: these most valuable of medicines are all the result of microbe life. We do not have to look far for evidence that microbes are good for us, because science has never managed to synthesize any drug as good as those that the microbe world produces. In some cases we modify antibiotics by making subtle alterations to the chemical structure, but we owe the antibiotics to our microbe cousins in the first place. The organisms are nothing special; most are found in the soil. If you visit a factory producing antibiotics, it has a distinctive odour of someone busy in the garden. The distinctive smell of freshly dug earth is produced by the microbes which hold the soil together. No form of life could seem more lowly, but it is from the tiny organisms beneath our feet that we obtain these vital and life-saving substances.

Meanwhile, microbes have germs of their own. We have already seen that bacteria are prey to a huge array of phage viruses. These are extraordinary entities, typically built like a polygonal box on top of a contractile stalk. The phage virus anchors itself to the cell wall of the host bacterium by means of a plate at the base and then, as the stalk contracts, the genetic material from within the box is injected into the cell. Once safely abroad, the virus genes take over the cell. All its energies are devoted to making more phage particles. The cell eventually ruptures, spilling out a cascade of new phages which can spread through the environment in search of new bacterial cells to invade. Viruses always seem malevolently disposed towards us, but here is a large group that are intent on destroying many disease-causing bacteria. The standard accounts say that phages were discovered by two bacteriologists, Frederick Twort (1877–1950) and Félix d'Hérelle (1873–1949). Twort was the first to spot them, whereas d'Hérelle (variously described as a qualified doctor and a bacteriologist) carried on the work. The phages were meanwhile ignored by the rest of science until the 1950s, when research began in earnest. That may be the standard view of the history of science, but it is very far from the truth. Twort was trained as a doctor at St Thomas' Hospital, London, where he graduated in 1900. In 1915, he was studying *Staphylococcus aureus* cultures when he noticed that some of them developed transparent holes where the organisms seemed to have been destroyed. Twort found that he could keep tiny samples of the liquid that remained, and transfer the same cell-dissolving property to other cultures of *S. aureus*. He realized that he had noticed an infection of the bacteria, although he did not carry the research on any further and always insisted that he deserved little credit for this observation. To him, it was an interesting observation that required further research.

Félix d'Hérelle was a very different character. He was completely untrained and, when he left school in Montréal aged about 15, he began an affair with a married woman. The estranged husband set out to find him, apparently with the police in tow, and d'Hérelle remained an outlaw for some time. He was fascinated by microbes but, rather than consult with bacteriologists, he decided to set up a laboratory in his bedroom and began to study on his own, at home. From there he worked in France and South America, and eventually held professorships at Yale and in Russia. In 1917 he made the same discovery as Twort, this time in enteric bacteria rather than staphylococci, apparently without knowing anything of Twort's precedent. Félix d'Hérelle, however, believed that this was a virus that could destroy pathogens, and he set up a huge campaign to promote the phage as a treatment of disease. He showed that the mystery virus could wipe out the bacteria from cases of dysentery in French soldiers in World War I, and embarked on an extensive programme of investigation. In 1921, he coined the term 'bacteriophage' (literally meaning *bacteria eater*) and published a book entitled *Le bactériophage, son rôle dans l'immunité.* Unlike Twort, who had moved on to other work, the bacteriophage dominated the rest of d'Hérelle's life. Although he is dismissed in most reference books, and the importance of his work on the phages is largely dismissed, d'Hérelle embarked upon one of the most prolific periods of study ever seen in the history of microbiology. Books poured from his desk, with handsome bindings and gold-blocked spines, and they told of thousands of different viruses. Everywhere he looked, he found more types of phage. He was invited to Russia, where he set up a highly influential centre to study and isolate phages. He became the world authority on this rapidly expanding field of interest.

In the end, his enthusiasms proved to be misplaced. The use d'Hérelle sought to make of phage viruses was as a treatment for bacterial infections. He collected and purified many phage types that could destroy intestinal bacteria, and some of them were used successfully to treat cases of dysentery, but as a rule the viruses proved unable to cure most bacterial diseases. Instead, the phages have had their main application in identifying different types of bacteria. Most of the major epidemics are caused by bacteria that we identify in this way (*E. coli* type O157-H7, which has caused fatal epidemics in recent years, is identified by its phage type, for instance). Since World War II, the only nation to continue collecting phages in case they could be used to treat bacterial infections is Russia, where there was a surviving laboratory established by Félix d'Hérelle, still run according to his teachings. After the fall of communism, the research ran into serious financial problems, and in 1997 a team from America were invited to the institute to discuss possible funding. The Russians were interviewed at length, and asked to give increasingly detailed accounts of their research; but in the end the Americans took their findings home for possible commercial exploitation and the Russians were left with no secrets, and no funding either.

As the phages are so hungry for bacteria, and can wipe them out so easily, why have they proved to be disappointing as agents of therapy? The reason lies in the difficulty of getting the phage viruses and the bacteria in contact. If they are injected, they tend to stay where they are rather than seeking out their prey. The viruses lack means of movement. They have worked in treating some cases of dysentery, but this apparent level of success results from two considerations: first, dysentery is usually a self-limiting disease and patients will recover spontaneously; and, second, the churning action of the gastrointestinal tract mixes the viruses and the bacteria and ensures that they come into contact with each other. Indeed, phage viruses in large concentrations still may have limited applications in cases such as this, but for most bacterial infections their potential is limited by the simple facts of nature. They certainly count as a friendly force, but we lack techniques to exploit fully what they could offer. When they do, the ghost of Félix d'Hérelle – untrained outlaw – can afford a smile at the scientific establishment's expense.

Phage viruses coexist with the bacteria in our intestines, and doubtless help to keep the community in equilibrium. Although we rely on our intestinal microbes in our normal lives, they are not absolutely vital for survival. We could certainly survive without them. This is not true of many other creatures, for there are innumerable animals that could not digest their diet without a permanent population of microorganisms. These tiny species work incessantly to recycle inedible components of the diet into proteins that the animals can digest. In ruminant animals (such as cattle), this process of digestion begins before the food intake ever reaches the stomach. Ruminants possess a large distension of the gullet, which bulges out to form a 'false stomach', the rumen. Inside this chambered pouch is a community of microbes which we still don't fully understand. It is quite clear that most of them are different from the types in our own digestive systems. There are cells of *Candida*, for example, but most of the microbes contain genes that are foreign to our internal populations. Instead, they code for enzymes that digest cellulose, and this is what allows cows to consume grass. Some of them are unfamiliar organisms, such as the crescent-shaped *Bacteroides succinogenes* and *Ruminococcus flavefaciens* which grows in delicate chains. Between them, they break down cellulose (which is otherwise impossible to digest) and produce succinate molecules for other bacteria to handle. Some bacteria then get inside the plant cells and break down the starches into sugar. Many of these are strange-looking organisms, too, like the large-celled *Epidinium* which swims actively with countless cilia thrashing in the surrounding liquid. Many species of rumen bacteria attack oils and fats stored within the plant cells, whereas others produce a range of products – glucose, acetic acid and a range of vitamins – which the cow needs for health.

The result of all this activity is that the bacteria grow in huge amounts. The ruminating animal regurgitates from its rumen from time to time, and stands around contentedly chewing the cud. This is the way it helps to grind up the plant matter and let the bacteria in to start their digestion of the food. All the while the bacteria are flourishing and reproducing as fast as they can. As the plant matter is reduced, the bacterial population increases. The result is that the vegetation is rapidly changed to a mass of bacterial cells, and they are rich in protein. Bacteria are not the end-product, however. The rumen is also teeming in larger protists: animal cells that swim around, grazing on the bacteria. These organisms are covered with a mat of beating cilia, and they waft the cells around like a super-efficient galley rowed by invisible slaves. These ciliated cells are stream-lined, to make them dynamically effective, and they scoop up bacteria into their funnel-like mouths much like the way a baleen whale sifts plankton out of the sea. The protists quickly increase in size, and when their volume has doubled they stop feeding for a while and divide in two. The two daughter cells swim off independently, and carry on grazing and feeding just as before. There have been many research programmes that investigate the bacteria of the rumen, and we understand them reasonably well; but protists need a much more complex environment in which to flourish and do not take well to growing in a Petri dish. Our fixation on molecular biol-ogy has given us greater insights into easily understood life forms, but we still know surprisingly little about the grander organisms that consume the bacteria. Some of them have astonishingly complex structures, with specialized regions within the cell to conduct nerve impulses and to move about. There are a few that lay down insoluble reserves inside the cell which look quite like features that you expect to see in more complex creatures.

Epidinium, for instance, has a segmented structure running down the cell, for all the world like the backbone of a conventional vertebrate.

As the rumen processes continue, these larger ciliates consume the bacteria and themselves become a huge, teeming community of single-celled animals. It is this vast swarm of mostly pure protein that is destined to pass into the true stomach, and which becomes the ruminant animals' true source of food. This paints a fascinating picture: cattle do not actually digest grass at all. Instead, they harbour a complex network of microbes that do the work for them. It is the rumen bacteria that actually digest the cellulose and turn it into proteins, vitamins and other valuable products. These bacteria are themselves consumed by the far larger protists, tiny independent animals that swim around collecting up the bacteria in their billions and growing as they do. It is these that pass into the stomach, and this is when the cow begins her own digestive processes. Clearly, she is not digesting vegetation at all – rather, she is digesting tiny animal cells. The cow, which seems to outsiders to be the most perfect example of a herbivorous creature, really digests nothing but animal protein. Yes, cows are really carnivores.

Wherever we look in the world of familiar animals, microbes are not far away. They play an equally important role in the life of vascular plants. In some cases, the existence of microbes on plants is already familiar. Silage is a fermented form of harvested vegetation, a kind of sweet sauerkraut, and is produced by the activities of the organisms that naturally occur on plants in the wild. The growth of the microbes means that silage is richer in protein than the raw grass, and the balance of population of microorganisms within the rumen changes to accommodate this altered diet. Wine is the result of fermenting grape juice, and the mature grape supports a surface layer containing yeast cells which will trigger the fermentation at the drop of a hat. You can see this layer with the unaided eye – it is the waxy bloom that marks out a fresh grape. In other plants there is a chamber containing a brew of microbes which help them digest their food. These types, like the pitcher plants, attract insects with a lure and trap them in a semi-liquid soup. Microorganisms living in this broth live on the decomposing insects, recycling their constituents to the plant.

But these are minor examples of co-operation between the world of visible plants and the microbial universe. The most powerful web of co-operation is hidden in the soil. We know that plants have roots that collect water and nutriment, and that most plants have root hairs, each a single cell thick, which give them intimate contact with the grains of soil and the film of moisture with which each is coated. It isn't true that all roots are covered with root hairs, mind; some (like the citrus family) lack them altogether. The root hairs are not the only point of contact between plant and the surrounding earth, however. Plants rely on microbes in the soil to feed them and protect them from disease.

In the autumn, Europeans traditionally go out into the forests to gather toadstools for a seasonal treat. We think of toadstools as wild plants which we pick like any other, but this is far from the truth. The toadstool is nothing more than the fruiting body of a fungus which rises above the earth to shed its spores into the breeze. The bulk of the fungus is the mycelium which lies hidden in the soil, like silvery traces of fine threads spread far and wide among the grains of the earth. What we see growing above ground is the tiniest proportion of the whole fungus body. The fungus in the soil spreads far and wide, interlaced with the mycelia of other species and growing round the roots of the forest plants in an intimate association. These fungi break down wastes in the soil and recycle them straight to the roots, ready to be absorbed. Many species of fungi go further, and actually penetrate the cells of the plant roots. They become a part of the host plant's structure. This intimate association between roots and fungi is known as a mycorrhiza. Some scientists write of a mycorrhiza as though it were a fungus, but that's not correct. The mycorrhiza is the name we give to the association between the plant and the fungus: it is the most intimate of associations between unrelated species, and supports the life of each. The fungus provides food materials for its plant partner, and the green plant channels some of its own products (manufactured with the aid of sunlight falling upon its leaves) down into the roots where they help to sustain the fungus. Some types of vascular plant manage without a mycorrhiza (they are said to be absent in the brassicas, for example) but for most they provide a vital life-line. You might offer an objection to this proposition: we can grow plants hydroponically, so this must be an improvement – and it also suggests that the fungi around the roots are not so necessary after all. The principle of hydroponics is that plants are grown on an inert substrate and are fed a mineral solution which provides everything they need. As these plants grow to be lush and verdant, microbes in the soil cannot be crucial. This misses the point. To feed the plant their mineral solution, we need to have a factory producing it. The consumption of energy in making plant fertilizers is always very high (it is rarely costed

The first record of parasitic fungi. In Hooke's *Micrographia* appeared several drawings of fungi. His portrayal of bread mould *Rhizopus* is unmistakable, and this figure shows his studies of a mould found on rose leaves. The organisms here are rust fungi. Although the details of emergence are not portrayed entirely correctly, the essential structure of the fungus is remarkably accurate.

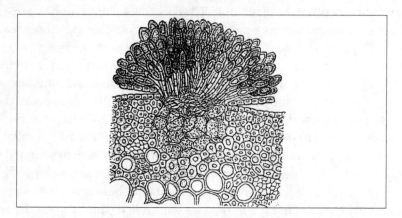

Victorian microscopists study the parasitic fungi. By the late nineteenth century, the rust fungi were well understood. These images, from Strasburger's great work on botany, show how the teleutospores burst out from the surface of cereals that they infect. The cereal rust has a complex life cycle, part of which must be spent infecting the barberry *Berberis*. Removing this alternative host breaks the cycle and prevents the disease in cereals, facts well understood by these Victorian microscopists.

into the efficiency of agriculture) and these factories are quite separate from the hydroponic units where the crop plants are grown. If you imagine the massive factory and the high-technology hydroponic establishment together, then you can imagine the massive investment of engineering, money and energy that a commercial hydroponic establishment truly needs to function.

Compare this with the situation in nature. Here there is little to see. The mycorrhizal associations are responsible for capturing and recycling minerals for the growth of forest plants, and they do it all without any industrial equipment and – most important – without any external supply of energy. This way of nourishing green plants is self-supporting and continually adapts to changing circumstances. Many modern townships are moving towards the composting of organic wastes, and this is clearly the way we need to move. Not only is there no need for external supplies of energy, but the whole system produces no hazardous pollutants and provides just what the host plant requires. Human technology is crude and consumptive by comparison.

The greatest value of fungi in nature is their role as recycling agents of dead matter and organic wastes. Relatively few of them infect humans. The little fungus *Trichophyton* lives in human skin, causing athlete's foot, ringworm or dhobi's itch (depending on where it occurs). Farmer's lung is caused by fungal spores in the lung. We have known that microscopic fungi infect plants since the seventeenth century, and it is now being proposed to use this to take action against the illicit growing of the opium poppy *Papaver somniferum*. There is a fungus that attacks this plant, *Pleospora papaveracea*, and Russian scientists have isolated a strain that could devastate the opium fields. Caution is the by-word, of course; we need to be sure that we are not releasing a new organism which could cause wider problems. The fungus is naturally found from Europe down to Australia, and it could become widespread.

Most fungi lie in wait as resting spores, only to germinate and grow into new colonies when there are suitable foodstuffs available. Others have sophisticated ways of entrapping their food, and some species have even developed traps which catch their prey. Predatory fungi in the soil produce large loops of hyphae, and when something swims into the loop the entire structure contracts in an instant to seize hold of the organism, which then dies and decomposes so that it may be absorbed by the fungus. Here too, at the microbe level, we find extremely complicated mechanisms, which are

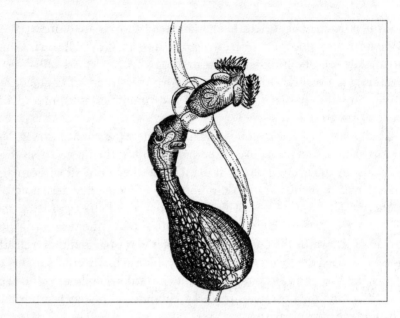

Trapped by a microscopic snare. Some microscopic fungi develop loops of hyphae which can ensnare passing organisms. The presence of prey makes the thin cells of the loop suddenly inflate, trapping the prey in a lasso. Digestive enzymes break down the prey, and the fungus absorbs the resulting nutriment. Here a rotifer, *Callidina*, has been ensnared by one of these highly developed predatory fungi.

redolent in many ways of the patterns of behaviour found in multicellular forms of life.

Acid rain is a popular topic of conversation, and schoolchildren are always being taught of its evils. The amount of pollution from industrial factories is indeed a hazard to the environment, and the amount of acidic fall-out has reached damaging proportions. Stone buildings are literally dissolving away, like a wedding cake left out in the rain. We should not lose sight of the importance of natural acids in rainfall, however. All rain is normally acid rain. It is meant to be. Among the components of rainfall are sulphurous and sulphuric acid, which sound like horrifying contaminants but which are actually vital for life. These are the principal source of the sulphates that upland grasses would otherwise lack. There is also nitric acid in rainfall, most of it resulting from volcanoes and discharges of lightning, and this is an important source of nitrates for green plants around the world. Plants have a fundamental need for nitrates, because these are the building blocks from which amino acids and proteins can be synthesized. The microbes in the soil recycle nitrates from dead organisms and supply them to the roots of growing green plants, and this is a crucial process for the growth of vegetation. The industrial production of nitrate fertilisers was first developed in 1909 by the German chemist Fritz Haber (1868–1934) who was awarded the Nobel prize for Chemistry in 1918. His process could make ammonia out of hydrogen and atmospheric nitrogen, and this was the ideal raw material for producing nitrogenous fertilizers. This allowed farmers to feed their plants with an abundance of nitrogen, and poor quality soils were brought into production as a result of this valuable breakthrough. The over-use of nitrates does need investigation, however, for the industrial production of these compounds imposes a burden on the environment which a less extravagant approach might allow us to control. Most of the air around us is nitrogen (four-fifths, 80 per cent, of the atmosphere is nitrogen gas). Some microbes have already conquered the problem of converting this harmless gas into the nitrates on which green plants can feed. This process has long been known as 'fixing' nitrogen, and the microbes that can perform the task are known as nitrogen-fixing organisms. The best known are the bacteria of the genus *Rhizobium* which infect the roots of leguminous plants – clover, peas, beans and pod-forming plants. The roots of these species develop warty growths, and look to the naked eye exactly as though they have contracted an infection. They have, of course; indeed newly germinated legume seeds soon become infected with *Rhizobium* and need this if they are to thrive. This explains why clover has long been grown as a way of improving soil. As the clover grows, the bacteria inside the root nodules fix nitrogen from the air and this nitrate is channelled into the plant as it produces new tissues. The rate at which nitrogen is fixed adds up to a surprising total; indeed there have been reports that half a tonne of nitrogen can be fixed each year in a single hectare. This is all the result of the bacteria that live inside the root nodules of leguminous plants, and is a remarkable total. Bearing in mind that this is all done without any external inputs of energy or chemical raw materials, it is a process that we need to value.

The importance of these plants was understood by the ancient Romans, who knew that clover could apparently miraculously improve poor soils. During the Middle Ages, crop rotation was widely practised as a means of boosting the output of fields. Every few years a meadow was left to lie fallow, and the growth of wild clovers boosted the nitrogen content of the soil ready for the next year's harvest. Although the value of these plants was clearly recognized, nobody knew what was happening until the 1880s. This was the decade when H. Hellriegel and H. Wilfarth grew peas and oats alongside each other on clean quartz sand. Both species failed to thrive. When the sand was inoculated with a small amount of soil from meadows where clover grew, the roots of the leguminous plants developed nodules and grew rapidly. The oats alongside, of course, failed to thrive just as before. In 1888 the Dutch chemist Martinus Beirjerinck (1851–1931) – whose name is often erroneously spelt without the first 'r' – discovered bacteria inside the nodules, and named them *Bacillus radicicola*. Since then we have recognized several species, and they are all grouped within the genus *Rhizobium*. Beirjerinck went on to recognize the existence of viruses for the very first time, yet all of his microbiological discoveries resulted from independent study as a side-line from his training in Delft as an industrial chemist. In 1892, T. Schloesing and E. Laurent grew leguminous plants in a closed chamber and ingeniously showed that the amount of nitrogen in the enclosed space went down as the amount in the plants increased. Here was the final proof that nitrogen was fixed by the plants, although it was not until 1952 that a radioisotope of nitrogen, $^{15}N_2$, was used to prove that the bacteria in the nodules performed the act of fixing the gas into a form that plants could use.

Although we always think of clover as the main nitrogen-fixing plant, the woody leguminous plants do so too. If you have a laburnum tree in the garden, then this is busy fixing nitrogen. Research in South Africa has shown that acacia trees can annually fix 200 kilograms of nitrogen per hectare (180 pounds per acre), which helps to nourish the entire ecosystem of the region. We now know that legumes are not the only plants to bear nitrogen-fixing nodules on their roots. These were first studied in the 1950s, and are now known as actinorrhiza. More than 300 species of non-leguminous plants have been found to rely on root nodules, and they range from members of the rose family to the alder. The growth of the alder tree *Alnus* has been believed to improve soil for a thousand years, and it has now been shown that a hectare of this tree – which few people would associate with nitrogen fixation – can fix over 100 kilograms (220 pounds) of nitrogen in a year. Some microscopic algae can fix nitrogen too. These are blue–green algae of the genus *Anaebena* and its close relatives. You can find cells of *Anaebena* growing among the tissues of the cycads from which sago is produced. The floating water fern *Azolla* has even developed special chambers in which *Anaebena* can live. *Azolla* is familiar in the West as a tiny ornamental plant, but the nitrogen-fixing abilities of its algal partner have made it an important plant to grow in rice-growing areas. It is even harvested as a forage crop for cattle in the tropics. Currently, there is interest in the leaf nodules that some plants possess, and it seems that there may yet prove to be nitrogen-fixing bacteria on the aerial parts of some important plants. Even lichens make a contribution. We think of lichens as crusty growths on wind-swept rocks. They are symbiotic associations of fungi and algae and many of the algae are able to fix atmospheric nitrogen. You will find lichens in forests of Douglas Fir in cool temperate zones, and it has been calculated that the algae in the lichens make an annual contribution of up to 10 kilograms per hectare (8 pounds per acre).

There are many nitrogen-fixing bacteria that live independently in the soil, and are not enclosed by a host plant. The first (and one of the most important) of these is *Azotobacter*, discovered by Beirjerinck in 1901; now we know of scores of genera and hundreds of species. They do not have such a profound input as the bacteria we have already mentioned, for their contribution to a hectare of uncultivated land is about 5 kilograms (11 pounds)of nitrogen per year. To this we must add the nitrate that is added to the soil by rainfall – much of which originates in lightning – because this can contribute a further annual total of 1 kilogram per hectare (12 ounces per acre). Plants derive plenty of nitrogen from natural sources – most of it deriving from microbes in the environment which are equipped with the right genes. Interestingly (for this should be a simple matter for geneticists) science has never managed to implant the genes that code for nitrogen fixation into some other species. The ambition of many agriculturists is to give cereal crops the ability to capture nitrogen from the air to help them survive in barren soil. Many have tried, and all have failed. This uncomplicated manoeuvre still eludes us, and the humble clover plant still steals a march on genetic engineering.

8

Engineering genes

YOUR GENES LURK DEEP inside the cells of your body. They are linked together like beads on a necklace, forming 46 thread-like structures called chromosomes. The genes carry the program that triggers this wonderful phenomenon of life. Each of your cells is the smallest thing that you can see. An ovum is easy to see; it is the size of the full stop at the end of this sentence. You can see amoebae with the naked eye, too; indeed the best way to collect amoebae is to let mud from the bottom of a pond settle out in a Petri dish, and then pick off the tiny pinkish or grey dots with a fine pipette. Your own cells are half the size of those amoebae, and you can just about make them out if you have excellent eyesight. Your body contains 100 trillion cells, and (with the sole exception of the red blood cells, erythrocytes) each normal cell in your body contains an identical set of the same genes. The number of chromosomes varies from one species to another, and is unrelated to how advanced or well developed the organism is. Goldfish have 94 chromosomes, dogs have 78; a cabbage plant has 18.

Chromosomes are made from DNA. The molecule of DNA is a little like a twisted ladder. Each side of the ladder is made from molecules of phosphate and sugar, while the rungs are made up of four different nitrogen bases (little units that contain atoms of nitrogen). These are adenine (A), thymine (T), cytosine (C) and guanine (G). They each meet together in the middle of a rung. Interestingly, there is a set order in which they will combine. Adenine only pairs with thymine, and cytosine only pairs with guanine. Each of these rungs contains two of the nitrogen bases, and is therefore known as a base-pair. All mammals – monkeys and mice, horses

The alphabet of life. These are the four different nitrogen bases: thymine (T), adenine (A), cytosine (C) and guanine (G). They are attached to the double helix of DNA in a precise order, and facing the same way. These comprise the nucleotides from which genes are made. Here we are encountering the four-letter alphabet that spells out the codes from which new proteins are made. Although four letters seem too few, a set of base-pairs just ten nucleotides long can be arranged in over a million different ways.

and hares, pumas and people – have about the same number of genes, adding up to 3 billion base-pairs. For comparison, *Drosophila*, the fruit fly has about 15,000–25,000 genes. Nematode roundworms have 12,000 genes and yeasts about 8,500 genes. *E. coli* has been thoroughly sequenced and we know that it has 3,237 genes. We share almost all our genes with apes, but if you were to look closely at the exact sequencing you would find differences in about five per cent of the genes. By the time you get to a mouse, there would be sequence differences in about 25 per cent of the genes – although a quarter of these differences would not be significant, because they would not materially alter the proteins that they produce. The genes within the human races are 99.9 per cent identical.

So the structure of DNA is rather like a simplified alphabet, with four letters instead of the 26 letters of the English alphabet. We have no difficulty in understanding that you can communicate the full flowery measure of the finest prose using 28 letters in various combinations, and there are alphabets elsewhere in the world with fewer letters than that (see page 43 in Chapter 2). It is only an extension of that idea to grasp that an alphabet of four letters (A, T, C and G) can write out the code which makes us essentially what we are. It is not the number of base-pairs that mark out the complexity of the information, but their abundance. The DNA in the average human chromosome contains about 150 million base-pairs, making a total of six billion base-pairs in each human nucleus.

Base-pairs are not the same as genes. It takes thousands of base-pairs to make up a single gene. The smallest genes contain a little over 1,000 base-pairs, and the most complex contain half a million. Each human nucleus contains about 80,000 genes, although in each cell only a small proportion of them are actually in use. The genes are responsible for coding the production of amino acids. Amino acids, when fitted together in a precise array, become proteins, and proteins are the soft and slimy material that makes up a living cell. The appearance and texture of protein are best illustrated by the raw white of an egg: it is translucent and slippery, gel-like and easily denatured by heat. The human proteins are made up from differing combinations of about 20 amino acids. The exact nature of the combination is very important. There are roughly 100,000 different types of protein in our bodies, and each one has to be exactly right. The proteins are an astonishingly varied lot, and range from the antibodies that attack germs to the haemoglobin in our red cells which carries oxygen. The enzymes that digest our food are proteins, so are the hormones that make us grow and perspire, develop sexually and feel anxious. The composition of proteins is a matter of importance, and many important diseases are the result of an unwanted alteration in protein structure. Since the nature of the DNA molecule was unravelled, it has been found that it takes three base-pairs to code for a single amino acid in the long assembly line. Thus, a gene containing 12,000 base-pairs will code for a product containing 4,000 amino acids, all assembled in the correct sequence. These have to be folded and linked together in a complex three-dimensional pattern to make a fully functioning protein. The DNA does not produce the proteins itself, however. Instead, it codes for the production of a smaller molecule which translates the code into a form that will act as a template for the production of proteins. This is messenger RNA (mRNA), a smaller molecule than the DNA that generates it. Using modern electron microscopes, the mRNA can be seen to be extruded from the DNA in chromosomes like branches radiating out from a fir tree, or the bristles on a brush. The DNA acts as a template for the mRNA molecules that are emitted from the chromosomes, except that a different base named uracil (U) replaces the thymine (T) of the DNA. Once the strands of mRNA are complete, they disengage from the DNA and move through the cell to the region where they are expected to initiate the manufacturing process. The mRNA is then read by the molecular systems in its own region of the cell, and the manufacture of new proteins is thus completed. We can read the code ourselves. For instance, the sequence cytosine–uracil–guanine (CUG) codes for the production of the amino acid leucine in the new protein. Similarly, we know that adenine–uracil–guanine (AUG) codes for the insertion of methionine. The genes are nothing more than a highly condensed reference library. When you bear in mind that there is all the information needed to code for a human being in the head of a sperm cell (which contains the smallest nucleus in existence), then you are conceiving of the ultimate microdot. The total length of DNA in each cell measures about two metres (about six feet), and DNA is truly the ultimate in miniaturization. It is hard to imagine any information system more condensed than that.

Alongside the active DNA in each chromosome is a lot of additional genetic material, and this controls the rate of production of mRNA and,

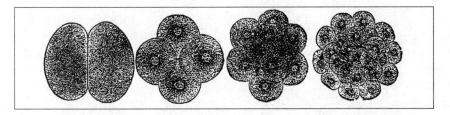

First divisions of the fertilized zygote. The new zygote divides into two cells, then successively into four, eight and sixteen. This well-observed study from 1864 shows that, as these early cells divide, they also become smaller. During these early stages, it is theoretically possible to separate the cells. Each one, under favourable circumstances, would have the capacity to develop into a complete adult. This is how identical twins, etc. begin life, and provides a mechanism for cloning.

indeed, whether it is time to make any or not. The regulation of genetic expression is a vital part of co-ordinating the way the cells work, and a large amount of each gene is taken up with these controlling functions. We now realize that the overwhelming bulk of the DNA in each nucleus never seems to perform any function at all. Geneticists call this 'junk DNA' and, although its existence has intrigued people for years, nobody is clear what it is there for. There is far more junk DNA than could possibly be concerned with regulating the way the genes behave, and it may be that this 'junk' is actually far more important than anyone has realized.

In some ways, the regulating functions of the genes are the key to the way life works. If DNA simply extruded mRNA like spaghetti from a pasta machine, then the cell would fill with all proteins at all times. Life proceeds only because the nature of the proteins, and the times when they are produced, fits a preordained pattern. The last thing you would want is a load of bright red haemoglobin produced by the cells that create the transparent lens in the eye, or masses of muscle proteins generated inside the nerve cells of the brain, which are meant to specialize in making you think. As it begins to develop, the freshly fertilized ovum (the zygote) divides up into smaller cells which are all virtually identical. The ball of cells soon becomes a hollow sphere, and then an elongated tube with a distinct head and tail end. As this phase unwinds, the multiplying cells begin to show differences as they

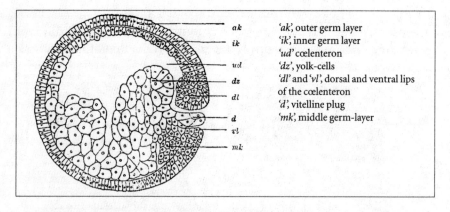

ak	'*ak*', outer germ layer
ik	'*ik*', inner germ layer
	'*ud*' cœlenteron
ud	'*dz*', yolk-cells
dz	'*dl*' and '*vl*', dorsal and ventral lips
dl	of the cœlenteron
	'*d*', vitelline plug
d	'*mk*', middle germ-layer
vl	
mk	

The newt embryo begins to specialize. The ball of identical cells (the blastula) soon begins to specialize into a hollow ball (the gastrula). By this time the tiny embryo has a distinct front and rear end, and a recognizable way up. The largest cells are many times greater in volume than the smallest. We can see further specialization in the figure on page 161.

start to specialize. In the terms I am explaining in this book, some of the multi-faceted abilities of an early cell dampen down while others are selectively potentiated. The regulators within DNA are the seat of this selective control. As a result of its subtlety, science has not yet come to grips with the details of the way that this system works.

We do, however, see examples of the illnesses that can result when things go wrong. If cells keep dividing when they ought to have stopped, tumours result. When proteins are folded wrongly, then they do not work as they should. The terrifying effects of spongiform diseases occur when the prion proteins within the brain cells fail to fold up properly, and holes develop inside the neurons because the protein molecules no longer fit into their proper place. In sickle-cell anaemia, a genetic disease found in Negroes, the haemoglobin molecules do not fold properly. As a result, when the haemoglobin is in its deoxygenated state, it tends to form long thin fibres rather than smooth folded bunches. The result is that the red cells, normally round and soft bags of gel shaped like car wheels, become elongated and pointed at the ends. This causes deep, boring pains in the joints and is a crippling burden throughout life. Interestingly, because the haemoglobin molecules are the wrong shape, they cannot be used as food source by the malaria parasite *Plasmodium*. People who inherit the sickle-cell gene do not suffer from malaria as a consequence, and in that sense gain an advantage over susceptible persons without the trait. There are about 4,000 inherited diseases known to medical science, and they confer a tragic burden upon families where they occur.

When cells divide, the chromosomes split in two so that a full set passes into the nucleus of each daughter cell. They were first seen by an extraordinary botanist, Karl Nägeli, in 1842. He watched dividing cells and wrote that tiny bodies were visible as the nuclei split in two. He called them 'transitory cytoblasts', but they were what we now call the chromosomes. The detailed stages of cell division were diligently recorded by Walther Flemming (1843–1905) who was a professor at Prague and later at Keil. He identified each stage of the process and gave them the names by which they are known to this day. As the 'transitory fibroblasts' stain easily with the microscope stains that we use to make specimens visible under the microscope, they were named chromosomes (from the Greek, meaning 'coloured bodies') by Waldeyer in 1888. Flemming made a crucial observation, and that was that the chromosomes split down the middle as cell division occurred. What happens is that the spiral structure of DNA unwinds like a nylon zipper, with new base-pairs being formed as each of the rungs of the ladder comes apart. Each chromosome splits into twins, and the double chromosomes become attached to opposite ends of the cell by fine strands like the cords on a parachute. These cords form the spindle of the cell, as it is called, and as they shorten they pull the newly divided chromosomes apart. They do not separate easily, and you can see that they need to be dragged away as though they were held together by some sort of microscopic Velcro. The spindles become shorter and thicker, the chromosomes are pulled apart, and eventually each of the newly divided sets is drawn to the two poles of the cell. A new nuclear membrane forms around each of these groups, as then the cell starts to narrow around the midpoint. The narrowing proceeds until the cell is nipped in half, and this is how one cell becomes two.

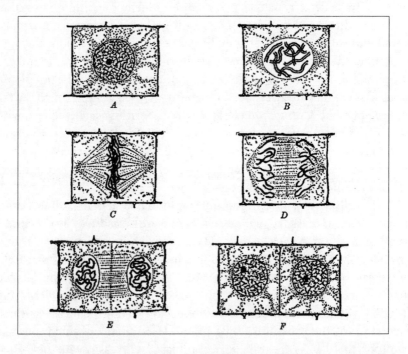

The crucial phenomenon of growth – the living cell divides. Edward Sinnott prepared these views of a dividing plant cell in the late 1940s. The normal nucleus (a) shows no sign of chromosomes. As division starts, they appear as paired threads (b) which congregate around the centre of the cell (c). As they are drawn apart (d) they form two daughter nuclei (e) and a new cell wall forms. Two new cells now exist, each with its own identical nucleus.

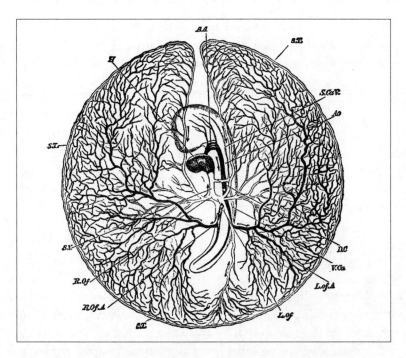

The early embryo establishes a network of blood vessels. Early in development the embryo establishes a network of blood vessels to bring it oxygen and nourishment, while removing wastes. This is a three-day-old chick embryo, illustrated in 1876. Here the connection is with the yolk. In humans, the network marks the beginning of the development of the placenta. Note the heart ('h') and aortic arch ('aa') which are already well formed.

The process is called mitosis, or binary fission, and is the key to understanding how organisms develop.

The story of the discovery of mitosis is an interesting aspect of the development of biology, and our knowledge of the structure of DNA allows us to see what is happening at the molecular level. However, there is one further mechanism that is crucial to cell biology. What happens in sexual reproduction? Each normal human cell contains 46 chromosomes, 23 from each parent. At the moment of cell division, the 46 split to produce 92 daughter chromosomes, and when these are pulled apart we are left with 46 per cell, just as before. Sexual reproduction involves a sperm and an ovum joining together to form a zygote, the fertilized cell from which the new individual will develop. If the germ cells were to contain 46 chromosomes each, we would end up with a zygote containing 92 chromosomes. At each generation, the number would double. After a thousand years, the weight of chromosomes in a single nucleus would outweigh the solar system, so clearly there must be some further mechanism at work. The truth of the matter is that, when the sex cells form, the number of chromosomes is halved. This is a different form of division known as meiosis, and it was unravelled by a distinguished English cytologist, Cyril Darlington (1903–81), at the independent John Innes Horticultural Institution. Building on the work of Thomas Hunt Morgan (see page 30–5), Darlington showed how genes may be crossed over from one chromosome to another, and elaborated the idea that the sexual division, meiosis, was a principal cause of genetic change between the generations.

What happens in meiosis is that two divisions happen, one after the other. The chromosomes congregate across the midpoint of the dividing cell and split in two, just as they do in normal mitosis. However, after the spindles have drawn the two new sets of chromosomes apart, there is a second division at right angles to the first. This time the chromosomes do not split in two, so half of the 46 are pulled into the two new nuclei. Thus, the sperm and ovum each contain only 23 chromosomes. These will be an assortment of the chromosomes from the parents, roughly half each. The cells of a woman contain two X chromosomes (they are known as XX cells) whereas those of a man have one X, and one Y (XY cells). Meiosis means that every ovum produced by a woman contains a single X chromosome, whereas of the sperm cells half contain the X and the other half contain the Y. This is why men confer gender on their offspring, and why the chances of having a boy or a girl are 50:50.

It is the genes that confer sexuality, not the chromosomes. In some individuals the male gene can be found on an X chromosome, and vice versa. Although they are exceedingly rare, there are cases of women with XY chromosomes, just as there are XX men. Some cases of intersexuality may be the result of a misplaced gene for gender, which ends up in the wrong place. Perhaps some of the controversial sports stars – who compete as women, but look like men – are innocent victims of a wandering gene. If this proves to be the case, a chromosome test will not determine the true sex of the individual: the sex chromosomes may be XX, as they are in a woman, but a male gene could still be present and alter the effective gender.

It is true that the way to tell the sex of a chromosome is to look inside its genes. Maleness and femaleness are not absolute characteristics. The human fetus is a bit of both, with sexual differentiation coming later.

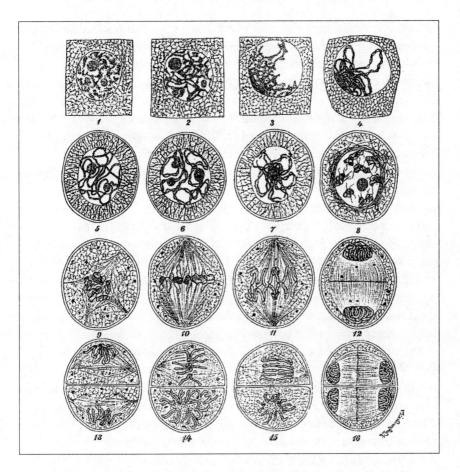

The double division that is the key to sexual reproduction. Sex cells need to have half the normal number of chromosomes. The total is restored when the two cells fuse and produce a zygote. Here we see a nineteenth-century study by Strasburger showing how this reduction division (meiosis) takes place. Normal cell division can be seen through to 12; but in 13 a second division starts. It takes place at right angles to the first (14–16) and consigns half the chromosomes to each new cell.

A newborn female child with an enlarged clitoris may sometimes be registered as a boy, just as a baby boy with undescended testicles and an abnormally small phallus can be believed to be a girl. The genitals of boys and girls are similar in many ways – indeed, men have a womb. This, the uterus masculinus, is a small fluid-filled body deep within the pubis and is homologous with the uterus of women.

In other animals, intersexuality is much more common. In goats, 15 per cent of some breeds of newborn goats are born with mixed male and female characteristics, and one pig in 500 is intersexed. A freemartin is a calf (usually the female twin of a bull calf) which is chromosomally a cow but has many of the characteristics of a bull. Chickens can apparently change sex. A mature fowl has one functioning ovary on the left, the other being shrunken to a rudimentary form. Should the left ovary be destroyed – by a tumour, for example – the right one develops to take its place. However, it matures into a testis, not an ovary, and at the next moult the hen produces the plumage of a cock. Although sex determination is the province of the genes, the expression of gender can depend on circumstance, and some fish naturally change sex during life. Parthenogenesis (virgin birth) is widespread in the aphids, which is why they manage to proliferate so quickly. All summer long, female greenfly produce their young without the intervention of a male. They abandon meiosis and their ova develop directly into the young. Sexual reproduction does not raise its head again until the autumn, when the need for explosive reproduction is over.

Our investigations of the chromosomes show that they are made up of three main parts. The DNA sequences are the main genetic component of the chromosomes, but capping each end of the chromosomes are telomeres, short sequences of DNA that are repeated many times over. The telomeres seem to be some kind of timing mechanism, as they shorten with each cell division. In the end, when it is time for an organism to end its life, the telomeres are so short that cell division slows up and the body dies. For all our medical knowledge, we have not increased the ultimate lifespan of humans, even though the average life expectancy has gone up. The telomeres are apparently nature's way of ensuring that each generation expires, making way for new descendants. Somewhere near the centre of the chromosomes lies a centromere, which is where the spindle attaches itself during cell division. We know very little about the centromere, which is currently a source of much investigation.

There is such complexity within the DNA molecules, and so much activity within the genes, that mistakes are bound to occur. Cells contain subtle systems in their genes which help to detect damage and correct it, but mistakes in the assembly of new base-pairs occur regularly. This is how mutations happen. As a rule, a mutation causes problems for the individual in which it occurs, but it also acts as a means of testing the cell and measuring how genetic changes behave. The tools available for examining DNA were readily available, but the problem was that the molecule itself is minute, highly complex, and was destroyed in the process of analysis. In 1972 an exciting breakthrough was announced by two American geneticists. These scientists, Herbert Boyer and Stanley Cohen, worked out how to amplify genes. They used restriction enzymes which cut out a section of DNA and recombined it with the DNA of another cell. Typically, they inserted their DNA sample into a rapidly growing bacterium such as *E. coli*, because this allowed them to produce a reliable supply of the DNA segment by growing the bacteria in culture. In this way, they could make billions of copies of a segment of DNA – more than enough to analyse and examine in great detail. If the segment contained the genes coding for the production of a hormone, such as insulin, then implanting this into the *E. coli* genome meant that you could grow the bacteria in tanks, harvesting the hormone just as you might produce alcohol from yeast. The technique is particularly useful for making supplies of antibodies. Genes responsible for antibody production are introduced into cells of *E. coli* and billions of identical versions of the antibodies are produced. The technique is a kind of cloning and, because they are all identical, they are known as monoclonal antibodies.

As the new discipline of recombinant DNA technology has developed, it has allowed us to develop rapid and reliable methods of mapping the exact sequence of base-pairs in segments of DNA, and allows us to determine the order in which amino acids are going to be produced by the resulting mRNA. We now know that the gene causing cystic fibrosis, for instance, has lost three bases out of the 250,000 in the entire gene. Thousands of human genes have already been mapped in this way, and the international Human Genome Project aims to complete the mapping of the genes in each human chromosome early in the third millennium. Sometimes we can identify healthy people who carry a mutated gene. These new techniques also allow us to alter the DNA at will. Then, if it was spliced back into the cell line from which it originated, you could have changed the genetic structure of the cells. In some cases the repaired DNA could be injected directly into the cell nuclei. Bone marrow cells (which produce the blood cells throughout the body) are one source of cells that could be removed, repaired and inserted again. Other diseases could be tackled differently, by splicing segments of DNA into a virus and then using the virus to carry the repaired DNA back into the body. Viruses are so adept at injecting DNA into the host cells that there is a most satisfying feeling when we harness the virus and use it to carry genes that will help to cure people, rather than making them ill.

There are two approaches that can be used. Sometimes it is possible to identify the rogue protein with the mutated structure, and then work back to the gene that codes for it. If this is impossible, then a positional technique can be used: the mutated gene is found first, either by finding the position of the similar gene in mice or by homing in directly on the structure of the gene. This technique is offering hope in the future treatment of such diseases as neurofibromatosis and Duchenne muscular dystrophy. Gene therapy may soon offer an answer to such crippling fatal conditions as cystic fibrosis and sickle-cell anaemia. There are also inherited diseases that cause breast cancer or dangerously high levels of cholesterol in the bloodstream which geneticists are investigating, because they may be amenable to gene therapy too.

We could cure many diseases by preventing the genes inside the cell from producing a harmful product. If we could switch off the gene, or stop its mRNA from carrying its information, the harmful product would be banished and the disease could be cured. There is an obvious model for this: the antibodies we encountered on page 111. An antibody is produced by cells under attack from an antigen – a protein that can damage them. The antibody is complementary to the antigen, and the two join together, thus inactivating the antigen. It is the antigens on the surface of disease-producing cells that trigger the production of antibodies. Some antibodies last virtually for a life-time (once you have had the infection, you never get it again). Others don't last long, so you need to have a booster from time to time. Many germs (such as cold and 'flu viruses) frequently change their surface markers, so that ready-made antibodies don't inactivate this new strain.

However, why not use the same system against genetic diseases? We say that a normal sequence of mRNA is constructed in the right 'sense' to work properly – but what happens if we were to make a strand of mRNA that was complementary in structure? This would be an 'antisense' strand. It would pair off with the 'sense' mRNA and neutralize it. All you have to do is produce a strand of RNA where everything is complementary:

    3'-C G U A-5'   sense mRNA
    5'-G C A U-3'   antisense mRNA

When the antisense mRNA joins with the normal strand, transcription is blocked. The two lock together, like antigen and antibody, and the unwanted messenger cannot exert its effects. We know it works, because of research to keep tomatoes fresh tasting. The first commercial application of this principle is the so-called Flavr Savr brand of tomato. Fresh tomatoes soon lose their flavour, because enzymes break down the compounds that make them taste so good. A shop tomato is usually much inferior to one picked from the vine. Transgenic tomatoes were introduced because they contained an extra gene. This produces antisense RNA which largely inactivates the enzymes, so the flavour is not destroyed so rapidly. The transgenic tomatoes have only 10 per cent of the enzyme, so their flavour lasts much longer.

There are hopes that the idea might be extended to the treatment of sick people. So far there have been mixed results from using antisense treatment for patients with lymphomas, but malignant melanoma has begun to provide some encouragement. It works like this. There are two markers, known as gangliosides, GM2 and GD2, which are found on the surface of cancer cells, including melanoma. GM2 acts as an antigen, and can stimulate the production of antibodies against GM2. It is almost as though you were immunizing a person against their own cancer cells. Research already suggests that patients with late melanoma can be helped by this treatment. To process the product, compounds from limpet blood are used. So far the 'vaccines' do not seem to cause any side effects, and the success rates are promising.

Antisense RNA occurs in nature, although we do not know how widespread it may be. There is a naturally occurring antisense gene which blocks a growth factor in humans and in mice. This is a gene coding for growth factor 2 receptor (Igf2r) and it is only inherited from the father (not from the mother). An antisense drug named GPI-2A is being tested in Canada as a treatment for AIDS. It blocks the replication of the virus and may stop transcription of the virus. The use of antisense therapy could have widespread applications in medicine, including Alzheimer's disease, cancer, heart disease and a range of inflammatory disorders such as arthritis. However, it is not as simple as it might seem. Some of the first trials were experiments with monkeys, and the antisense drug seemed to work as it should – but the animals all suffered from heart attacks. Nobody expected this, and it reminds us that ideas that look good on paper rarely translate too easily to clinical practice.

Future research into such possibilities will be helped by technology. You can now buy machines that construct experimental DNA to order. You enter the base-pairs by pressing buttons, confirm the entry which appears on an LCD digital screen, and the hardware produces a strand of DNA with exactly the sequence you chose. It is hoped to create healthy genes that could replace disease-causing genes. However (as the unexpected problems with antisense mRNA proved), we need to know a lot more about the operation of these substances before we start inserting synthetic genes into cells.

One far-reaching prospect has been the creation of complete new chromosomes that could carry cures directly into cells. The first artificial chromosomes were made for yeast cells in 1987. Now, the first self-assembly chromosome for human cells has recently been made by Huntington F. Willard of Case Western Reserve University School of Medicine in Cleveland, Ohio. They put three segments of DNA together: some of the normal genetic DNA, some DNA from the telomeres, and some of the alpha-satellite DNA which is believed to be crucial for the functioning of those mysterious centromeres. The components assembled themselves into a new chromosome. This was injected into cells in tissue culture, and kept going through many successive cell divisions for six months. The study of these synthetic chromosomes offers a unique chance to study the details of how genes function, and this model is one of the most promising techniques for getting to grips with the regulation of genes. They are unlikely to be used in gene therapy for a long time, because they are too large to be transferred by viruses and we do not have any other means of introducing them into the cells.

Many experiments have shown that human genes can survive inside the nuclei of experimental mice, but these experiments usually transferred no more DNA than about 200 genes at a time. In 1997 entire human chromosomes were inserted into the cells of mice, who have 40 chromosomes. The animals carried on apparently as though nothing had happened. The research was carried out by Kazuma Tomizuka of the Kirin Brewery Technology Laboratory in Yokohama, Japan. They showed not only that human chromosomes could add themselves to those of mice, but that the mice could grow and carry on without untoward effects. These chromosomes each contained 1,000 genes, 20 times as many as had ever been transferred before. As they are genes from humans, they seem to be ignored by the mouse cell. They did show the proper responses of genes, however. Thus, the genes that are supposed to be active in specific areas of the body did indeed turn themselves on and off alongside the genes of the mice, so it is clear that chromosomes can function in a different species. Many illnesses (such as Kawasaki's disease) might be cured by antibodies, and an American company, Cell Genesys of Foster City, California, have now produced human antibodies in laboratory mice.

The realization that genes occur within the chromosomes is crucial to our understanding of multicellular organisms, but is not the whole story. There are other genes within cells, genes that lie outside the nucleus and do not take a part in mitosis. Scattered through the cytoplasm are sausage-shaped bodies, the mitochondria, where the energy of the cell is released. As we have seen (page 57), mitochondria are the cell's power house. The idea has existed for over a century that mitochondria began as independent bacteria, and took up residence inside cells as evolution proceeded. It has also long been believed that green plants acquired their cholorophyll-containing plastids in the form of free-living, independent algae. If this is the case then you would expect that the free-living antecedents of these organisms would need their own DNA, and it is not surprising that we find DNA in mitochondria and plastids.

If you look at a single-celled plant such as the alga *Chlamydomonas* (which contains a single plastid) there are 500–1,500 DNA molecules adding up to 15 per cent of the cell's DNA content. The leaf cells of the garden beet contains 35–45 plastids per cell, and about 100 tiny chromosomes within each plastid. In total, there are from 1,000 to 6,000 DNA molecules inside the plastids of each cell. Yeast cells contain about 30 mitochondria containing up to 5,000 genes, whereas in humans there can be 1,000 mitochondria containing maternal DNA in each active cell. These genes do not follow the laws of inheritance. They do not line up to take part in cell divisions, and indeed the number of chromosomes per plastid (or per mitochondrion) does not remain constant. The cytoplasm of the ovum contains large numbers of mitochondria, so the total amount of DNA within a fertilized egg cell is always more than the DNA contributed by the chromosomes from both parents. This free DNA comes from the mother. One of the tenets of Darwinism and the laws of inheritance passed down by Mendel is that hereditable characteristics must be passed on without any contribution from the life-time of the parents. This was the answer to the ideas proposed by Lamarck, who thought that giraffes obtained their long necks by stretching, and passed on the trait to their young. The idea of survival of the fittest holds that the long neck develops over the generations, by the death of animals with shorter necks so that those that can reach higher survive longer, and breed more. The passing of DNA in the cytoplasm of the ovum opens up a range of different possibilities, and could certainly imply that some characteristics acquired during life could be transferred to a new generation. Some hereditable conditions (such as muscular dystrophy) are linked to cytoplasmic DNA and others are likely to be discovered.

The research that will reveal exactly how great is our problem in understanding human genetics is the Human Genome Project. Planning started in 1986, and the programme was launched in 1990, as a co-ordinated effort to identify all the estimated 80,000 genes in human DNA. Scientists will have work out the sequences of the three billion bases in our DNA, and also have the brief to consider the ethical framework of such genetic research. The DNA samples come from many sources, so the resulting sequences will not represent a single person. The most complete map of the human genome, published in 1997, featured about 8,000 landmarks (twice as many as ever before), including some parts of chromosomes that are known in considerable detail. So far, less than 10 per cent of the genome has been sequenced, because research has originally centred on developing reliable and rapid methods of sequencing rather than concentrating on mapping the genome itself, but now the mapping itself is rapidly proceeding. Scientists in about 50 countries are participating, including Australia, Brazil, Canada, China, Denmark, the European Union, France, Germany, Israel, Italy, Japan, Korea, Mexico, the Netherlands, Russia, Sweden, the UK and the USA.

The unravelling of human genes is clearly a project that could open our eyes to new ways of treating tragic illnesses, yet it brings with it problems that we have never faced before. One of them is the responsibility that our new knowledge implies. Once we know about our genes we may well have simple tests – like the pregnancy test kits of today – which can screen for specific genes. Naturally, insurance companies base their calculations on demonstrable risks (or good guess work, if there is little precedent on which to base calculations). If you turn out to harbour the gene for some condition that could shorten your life, or damage your health, then a company would quote you a far higher premium than it would for someone without that handicap. To demand that people are tested would be an infringement of liberty, but nobody could stop a company from offering a discount to people with a clean record of tests.

There is already a precedent for this attitude. Since the mid-1980s, life assurance companies have asked whether an applicant had ever taken a test for AIDS. It always struck me as an interesting question: they were not asking whether you were HIV positive, but whether you had taken the test. Someone careless in their sexual habits – who was at risk, but would not think to be tested – was clearly cleared by this question. Someone who was highly responsible, and had wished to be tested as a wise precaution, would have failed the insurers' test. Similar questions on genetic testing may yet appear on insurance proposals. A recent survey of the top 500 American companies suggested that only three per cent were interested in the idea of a genetic screening for new employees, but that reflects only the current position. In future, as attention focuses more on these matters, the interest is going to increase. We are faced with some paradoxical situations: under British law an employer cannot fire an employee for a genetic handicap, nor can they discriminate against a handicapped individual on the basis of their disability. However, were an employer to discriminate against an applicant because of a genetic test which indicated a susceptibility to a disease, they would be in the clear. In today's world, you can be legally debarred from employment on the basis of a condition from which you do not yet suffer – but could not be dismissed if you already had it. This matter needs to be urgently addressed, because it could impose an appalling burden on people who are already facing problems that many of us would find unendurable.

What troubles most people, though, is the widely publicized idea of creating new species. After all, if we have managed to implant human genes into other organisms, could we not make completely new, unnatural species? We have been doing that for thousands of years, and many of the species are among the most familiar forms of life. There is nothing new about creating novel animals and plants, and nobody should be too troubled by the prospect. Sheep and goats, horses and cattle, dogs and cats are all animals created by human ingenuity, and are far removed from their wild relatives. The dog is *Canis familiaris*, a new species created by our prehistoric ancestors. Nobody knows exactly how they did it.

Much the same can be said of the plants that we use for our survival. Wheat and barley, oats and rye are all species created by humans. Some of the important crop plants, including maize and bananas, cannot even survive without human support. Some breeds of dogs, pigeons and goldfish are monstrously disfigured. Those genetic innovations were done by crossbreeding, and now we can identify genes and transfer them from one organism to another on a scientific basis. Surprising as it seems, the results

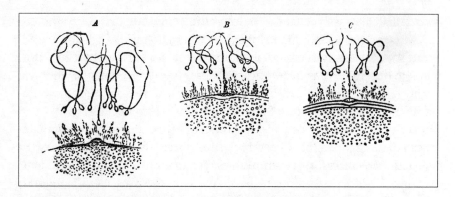

When new life forms – the moment of fertilization. As spermatozoa cluster around the ovum, a single sperm cell penetrates, shedding its tail as it enters. These remarkable studies were made in 1877 and show the moment that a starfish sperm cell is received by the egg. Modern techniques sometimes implant a human sperm head directly into an ovum. This eliminates the race between all the millions of sperm, and we need proof that there are no deleterious effects on the resulting baby.

of the work of today's genetic engineers have been unimpressive by comparison. They include a red petunia and an extra-productive strain of rice. For all our scientific knowledge, we have done nothing as remarkable as the majestic achievements of those forgotten, prehistoric experimenters who created new species, and then perfected what they wanted.

The ease with which we can handle living cells and grow them in the laboratory is astonishing, and it dates back earlier than you might expect. The transferring of embryos between one mother and another seems a very modern technique, but was first successfully achieved over a century ago. The experiment was successfully carried out by Walter Heape, a businessman who learnt embryology in his middle years, and became a highly proficient research scientist. He had a successful career in North Africa, Australia and New Zealand dealing in rice, sugar, milling and textiles. In 1878 his health failed, and he gave up the world of business to study histology and botany. By 1882, he was a Demonstrator in Morphology at Cambridge University, and contributed to Foster and Balfour's textbook *Elements of Embryology*. After a spell at the Marine Biological Association at Plymouth he went to explore the Himalayas, where he became seriously ill with 'acute rheumatism and fever'. He studied monkeys, paying special attention to menstruation, and from this moved to study menstruation in women from which arose his interest in fertilization. On April 27 1890, he took new embryos from the uterus of an angora doe rabbit. Each fertilized cell had divided into four, and he could clearly see this under the microscope. Meanwhile he'd had a Belgian hare doe mate with a male of her own breed, and he introduced two of the tiny angora embryos into the receptive prepared uterus.

The result was a triumph for his skill. The Belgian hare gave birth to six young. Four of them were clearly of the same breed as herself and her mate, Heape wrote, whereas two of the young were angoras. Not only was the hair recognizable, but a behavioural characteristic was evident, too: the angoras have a habit of swaying their heads from side to side. This was also seen in the young from the Belgian hare, and was the final confirmation that his embryo transplants had worked. His paper was presented to the Royal Society in November 1890, and was not even sent out for referees to check first. By 1906, Heape had published a book entitled *The Breeding Industry*. In it he set out a criticism of government intervention in science, which has a distinctly present-day ring to it: 'It is clear that at present a government department stolidly blocks the plainly defined road of progress,' he wrote. 'If such obstruction is to be removed, it is also clear that the board of Agriculture and Fisheries must be reorganized on a broad scientific basis.' Heape carried out his experiments with the help of Samuel Buckley, a well-known Manchester surgeon, and applied for his Royal Society grant from an address in Prestwich, Manchester. It is a quirk of fate that led to the world's first successful embryo transfer in a human being carried out in the same area. Seventy years later, the reproductive biologist Robert Edwards from Cambridge teamed up with Patrick Steptoe, an eminent surgeon from Oldham, Lancashire. Their techniques, now somewhat refined by decades of research, led to the birth of Louise Brown, the first test-tube baby.

You might wonder why an implanted embryo is not the subject of a massive immune response, but that argument applies as much to a normal pregnancy. If you are a pregnant reader, you are harbouring an individual with a very different genome to your own. If tissues from your child were to be implanted into you once the child is born, your own body would react violently to those tissues, recognizing them as completely foreign. During pregnancy, however, this alien occupant is tolerated and no immune response is forthcoming. The mother's body does not respond to an embryo in the uterus, which is why Heape's experiment worked so well.

Since this time we have shown how to fertilize ova and then implant them, and we have also found out how to culture cells in bottles and flasks.

The first great success for cultivating human cells was with specimens of tissue taken from a tumour in the body of a black American woman, Henrietta Lacks, in 1957. The cells took well to growing artificially, and have been kept going ever since. This cell line has adapted to an artificial environment, and now – known as HeLa cells for short – there are tonnes of them in culture all around the world. Ms Lacks herself eventually died of old age, but her cells live on.

Science has used *in vitro* techniques to give parenthood to couples who were previously barren. It can be a wonderful use of scientific research, although some cases require thoughtful appraisal. One recent headline-catching application of the technique was the pregnancy of Diane Blood, who was fertilized by her husband's sperm three years after his death from meningitis. As he had been comatose at the time she requested that the sperm be saved (and was in no position to give written consent, as required by British law), she had to apply for an export licence for the sperm samples so that she could be inseminated in Belgium. We need to address the emotional and psychological impact of knowing the circumstances of an unnatural conception.

The earliest method brought into use was AID, artificial insemination by donor. This gave infertile men the chance to see their wives pregnant by the sperm of another man and giving birth naturally. Women with medical conditions that prevent normal fertility can have a donor egg fertilized by the partner's sperm and implanted in her uterus to develop normally. There is a new technique for fertilizing eggs, which takes the head of a single sperm from the testicle and injects it into the ovum. It seems exciting and obvious that many people have been celebrating its success, though we need to think through these techniques carefully before we introduce them as routine.

In conventional reproduction, a single sperm manages to penetrate the ovum after a vast community of millions of sperm fights its way to its target. There are plenty of opportunities here for the fittest sperm to succeed, and it seems obvious that this is a selection process. If we take a sperm nucleus directly from a testis and inject it straight into the ovum, any process of selection is circumvented. Could this have deleterious consequences for the embryo? Perhaps so: Australian research has already claimed that tests of such children at one year suggest they may be significantly less mentally able, on average, than children conceived conventionally. The lesson is clear. Circumventing the ways of nature may bring unforeseen problems in its wake. Rather than being carried away by a heady sense of novelty, scientists need to be self-critical and cautious before introducing new measures into daily use.

As mentioned previously, the first production of a baby from *in vitro* fertilization was Louise Brown, born in England in 1978. A more recent candidate for international publicity was Dolly the sheep, who popularized the novel idea of cloning. Or is it so new? Cloning has long been used to produce plant crops, and all human identical twins are clones. As we have seen, the genes in cells throughout the body are the same, even if some of them are expressed at different times in different cells (or even not expressed at all). You should be able to remove the nucleus from a zygote, change it for a nucleus from another animal, and then have the zygote develop in its new guise.

The first mammal to be artificially cloned, Dolly the sheep, was produced at the Roslin Institute in Scotland during 1997. The nucleus was taken from the mammary gland tissue of the nucleus donor, and the image of breasts brought the well-endowed country singer Dolly Parton to the minds of the researchers, which is why Dolly became the obvious name for the lamb that resulted. In 1998 the first reports were released of cloned mice, this time from America. Cloning has long happened in humans, but usually our cloned cells are produced simultaneously, and without the need for intervention by a scientist. We call the result twins (or triplets, or whatever). They occur if the two cells, which form as the zygote first divides, drift apart and start developing into separate embryos. These are clones, and there is no mystery about that. It has been known for generations that you can do the same artificially with frogs and toads. Aphids manage their explosive reproduction through cloning, rather than through sex.

Were we to separate the cells of a new human embryo growing in the fertilization laboratory, I dare say that you could produce many identical humans. It has been theoretically possible for years. The technique has long been used in the growing of plants. Pineapples, for example, and orchids can be produced artificially by separating out cells from a single plant body and turning out identical progeny. Apart from its value in commercial agriculture, the technique is of particular value in creating large numbers of new plants from a species whose survival might otherwise be threatened.

The idea of making identical humans is an old one. The earliest novel on the subject was A.E. van Vogt's 1945 book *The World of Null-A*, and the first book with 'clone' in the title was P.T. Olemy's *The Clones*, published back in 1968. More famous was Ira Levin's story *The Boys from Brazil* of 1976, in which a vengeful scientist produced more little Hitlers than you'd find in the tax office. It is not so easy to produce a posse of malevolent dictators. What made Hitler into Hitler was not just his genes, but an unrepeatable series of experiential inputs which honed his personality, and the quirky use of a razor.

Even identical twins are far from identical. They may look much the same, but the details of the pathways of blood vessels vary greatly. Under a microscope, you can see that the way nerve cells join up varies enormously from one individual to another, and it is in complex systems like this that the personality is refined. Anyway, you don't clone babies. What you do is clone someone who becomes an angry adult with an attitude and a big stick. We already implant embryos of alien genetic origins into infertile women, and that is potentially a great cause of future problems. When you choose a partner, you do so because of personality attraction – but with a donor gamete you might end up bearing somebody from whom you are instinctively alienated. The problems facing us from in vitro fertilization of donated sex cells may make the controversy over cloning seem modest by comparison. Not only that, but cloning is not very reliable. Before Dolly was successfully delivered, there were 237 attempts to make the experiment work.

No sooner had the results been announced than a 69-year-old physicist from Chicago announced that he was planning to set up the world's first cloning company for humans. Richard Seed plans to call it the Human Clone Clinic, and he knows that desperate people will want to try the process. In my view, there might be cases where people will plead for cloning – when a child is lost, for example, and a parent is smitten with infertility. People like Dr Seed will be seen as a saviour for a lost cause. If that is true of the desperate few, it is not the case for the majority. After the idea was first announced, there was a news poll carried out in the USA by the television company ABC. Over 16,000 people responded to the questionnaire, and three-quarters of them were opposed to the cloning of people. Public opinion isn't necessarily the right arbiter, of course (there were plenty opposed to electricity or anaesthesia when they were new). Cloning of humans is now possible and, if we do have a new technique that could be of benefit to suffering humanity, it would be difficult simply to ban it in cases where it was desperately required. The loss of a child or medical treatment of a partner which prevented their fathering children could make cloning the only way of relieving childlessness. It is a twist of fate that the chief proponent of cloning, Dick Seed, sounds like a well-tried remedy for barrenness. 'This will bring people closer to God,' he says. I don't know about that, but it might bring Dr Seed closer to a new Porsche.

What disturbs people about ideas like genetic engineering and cloning is their newness, rather anything inherent in the nature of the processes. Garden plants, farm crops and domestic animals are far stranger creations than anything that geneticists have attained. If geneticists had produced a pekinese or a cauliflower the media would be up in arms. People are currently campaigning against cloning and genetic engineering, as though it defeats our essential humanity. We should campaign against the hasty introduction or the misuse of any of these techniques, but campaigning against the processes themselves is fruitless. They exist, and the need is to ensure that they are harnessed for our collective good and not for short-term commercial advantage. A friend, reading these words as the book was being written, remarked that he had attended a conference on genetically modified organisms (GMOs) in which a speaker was talking about the release of transgenic plants into the environment. When asked what they would do if something went wrong, the speaker apparently said: 'No problem, we'll just collect them up.' Such talk is dangerous. Plants may be easy to recover, but escaped genes are not. Some current teams are developing crop plants that are genetically modified in order to be resistant to patented herbicides. The two can be sold together. This trend is highly questionable, and we need to debate the issues in far more detail.

It is quite clear that genetically modified organisms pose specific problems. The official controls around test sites are insufficient to guarantee that new genes do not spread. Engineered genes could enter populations of wild plants, and that could cause us long-term problems. Marker genes for antibiotic resistance are easy to test for, and are often used to confirm the transfer of genes to engineered crops. Antibiotic resistance could be spread to wild bacteria, increasing problems for the future. The use of blanket weed killers could threaten survival of wild birds and mammals, and also have severe effects on microbial populations of the soil.

Other objections are not so sustainable. Some strains of resistant crops are grown by organic farmers because their resistance is thought to be natural and safer than a chemical spray. They may be natural, but they poison pests right enough, and little is known about their effects on us. Plants are potent sources of compounds that cause cancer and mutations, and the fact that there is something that wipes out insects hidden inside your lettuce does not make it inherently safer than something you buy from a shop.

The argument that all scientific measures are 'unnatural' holds no water. Towns and cities, doctors and nurses, schools and holiday resorts are all completely unnatural. Bread, butter and cheese are all the products of technology, and are utterly unnatural. We already accept highly unnatural means of altering our destiny. The use of donor eggs and sperm is regularly touted as giving people a chance to have children of their own, but this is not the truth of the matter. Children conceived and regarded as the product of a pair of people do not share the same genes that they would have shared if the cells had come from their parents, rather than from a donor. The most recent area of research focuses not on providing a donor ovum, but on using the cytoplasm from a young ovum to replenish the ageing ovum of the biological mother. This is certainly more natural, because it means that the new child inherits the mother's genes. In my view it remains a questionable procedure, because we are transferring some genetic material from the donor egg to the maternal cell. Some diseases (such as muscular dystrophy) are linked to cytoplasmic DNA. Adoption, where a child is taken to be the same as one biologically conceived, is fraught with potential problems. Alien genes are clearly likely to express themselves and, once in a while, a behavioural characteristic that is anathema to the parents must surely become expressed in the child. This may well account for those cases where adopted children are suddenly rejected as they begin to mature. We like to believe that a child becomes what its parents teach it, but that is not the case. Teaching can potentiate a child's genetic nature, and what a child learns dictates much of what it does as an adult, but this cannot deny the genetic nature of the organism. We are what we learn, but we become what our genes let us become. What matters is not nature versus nurture, but the nurture of nature.

We are already on the trail of genes for specific characteristics. There is now known to be a gene that regulates an enzyme that boosts the transfer of oxygen and nourishment to muscle cells. This is angiotensin-converting enzyme (ACE) and the ACE gene may explain why some people are better athletes than others given equivalent training. I fully expect us to find genes for musical, artistic and mathematical abilities. There must be genes for a whole litany of potential within us, good and bad. Just because people do have genes for specific characteristics, it does not mean that they are destined to follow that course. We are cognate beings with high intelligence, and our conscious actions are governed by decision-making processes which make us aware of the consequences. Meanwhile, we are handicapped by our ignorance.

For example, fat people often feel that a tendency to obesity is all in their glands, but they are regularly told that they are overweight as a result of gluttony. In my view, there must be a kind of 'adipostat' (a fat-controlling centre in the body) which regulates our adult body weight. It would be absurd to pretend that we have to control our weight by a careful balancing of everything we eat, because no other species counts calories and joules on a wall chart. As I have pointed out in the past, a person who gains just five grams a day (one-sixth of an ounce) would weigh close to 15 tonnes in old age. All species have genes that regulate their size, which is why there is no such thing as a five-kilogram sparrow. There are many reasons why people would benefit from being fatter than average. In biological terms, a person with extra fat reserves is better placed to resist the effects of a wasting disease, for instance. As size is a characteristic of all species, it must lie in the genes and this is why I am convinced that we will discover genes for fatness. You can limit the amount of weight gain by controlling the amount of fat intake, but it must be very galling for two people who eat similar diets, when one ends up as thin as a rake and another becomes far heavier. The need to accept our nature is crucial, because the pressures of modern communication make people aspire to impossible ends. Many young men are frightened of cholesterol, heedless of the fact that is insulates the nerves and is vital for life. Young women are taught to fear becoming fat by tabloid propaganda (one recent picture of a shapely young actress described her as 'chubby') and some become anorectic. An understanding of the diversity of human life is crucial for the future. People die through a mismatch of expectation with reality.

American agribusiness was quick to latch onto the potential of recombinant techniques. A gene that makes plants resistant to a common herbicide (glyphosphate) has been transferred to crop plants. This means that you can plant fields of the transgenic crops, spray the glyphosphate over them all, and you kill every plant except the crop, giving the farmer a monoculture of the chosen plant. But wait a moment: do farmers need this?

Are farmers currently unable to grow their crops? Is it best to have fields in which all plant life is destroyed – apart from a single crop – and are we so desperately short of certain crops that we need transgenic modification? Agriculture that does not heed the language of nature ruins the land. A huge monoculture would probably make it easier to mechanize farming, but that pays little heed to the long-term nurturing of the landscape. Growing plants in sterile soil, feeding them on fertilizers and dosing them with herbicides is nothing more than hydroponics, where you don't really need soil at all. There is a great microbial community in the earth (which few of us ever see, and none of us understand) and it is all at risk.

I am equally concerned about spreading a gene that confers resistance. Plants that grow in nature are well spread out and form mixed communities. Pests have adapted to survive the journey from plant to plant. In a farmer's field, there are great numbers of the same plant standing side by side. A pest can spread from one plant to the next at amazing speed, and the use of a pesticide spray can be vitally important if a pest is to be stopped in its tracks. We already have enough problems from bacteria that are travelling the world, picking up new resistance genes that help them resist our antibiotics. Knowing how plants can cross-fertilize makes me concerned that we are conferring resistance on crops. Resistance genes are best kept to ourselves, and not spread into the opportunist communities that coexist with us.

Could it happen? The scientific evidence for risk exists. There is a research programme into glyphosate-resistant sugar beet. Beet is one of the species already known to cross-pollinate wild beet growing nearby. The new genes could be transferred in the process. This could provide an avenue for the acquisition of transgenic resistance by wild plants.

There is a subtler problem looming up as well. I wrote above '… you can plant fields of the transgenic crops …'. That is not true for everybody. You would have to buy the crop from the producer with the patent, and pay for the rights to grow it. You would also have to buy the specified herbicide. In short, farmers round the world would become locked into a single licensed source of staple crops. There would be little hope of saving seed and using it next year. Genetic engineers have perfected a way of killing the seed produced by transgenic crops, a new invention known as terminator technology. A set of genes is known to be crucially important as a seed develops, and these 'late embryo abundant' genes, lea, can be knocked out by a set of repressors that are inserted into the genome of the crop. This means that the farmer buys viable seeds, but the seeds those plants produce cannot mature.

Were all these manipulations to become widespread, world agriculture would be subject to the demands of multinational corporations. You may say that it's up to the farmers to choose what they buy, and I hope that we always remember the force of personal choice when it comes to choosing a supplier. What concerns me is the message that the large multinationals keep giving out. One spokesman said: 'Within five years, all major crop plants will be transgenic.' Another tells me that the byword is now 'planning for profit'. In my view it is high time we began planning for people.

As a result of the sense of excitement in genetics, the companies seeking to exploit these ideas shoot ahead on the stock market more than most other sectors. Much of this is on the basis of promise, rather than actuality. The list of achievements is so far surprisingly disappointing. A version of insulin produced by genetically modified *E. coli* had to be withdrawn from the market, because it was not as safe as more traditional forms of the drug. In today's world, pure science has become impure. The demands of investors and the need to find funding based on results makes scientists into salesmen. I know that scientists tell all their friends that they hate publicity and are embarrassed at appearing in the newspapers. In the real world, scientists court publicity at every turn. Their success is measured as much in column inches as it is by publication in the scientific journals.

Newspapers are always hungry for a story, and something scientific often satisfies the demand. It may be incomprehensible. The tale may be told in long words (no self-respecting scientist ever uses a simple word, like 'test', where there is a polysyllabic alternative, such as 'quantitative analytical procedure'). But it is going to change the world, and that's what counts. There was an example recently in the news. At the Max Planck Institute in Germany they have been growing rat nerve cells in culture dishes. They put a microchip in the bottom of the dish to record the electrical signals given off by the nerve cells as they grew. This is not hard; anyone could do it. Nor is it new – there are dozens of laboratories around the world doing just that. So why has the Max Planck Institute seized the attention? That's easy to answer. The official spokesman claimed that this shows that we could 'connect the human brain to a computer'. This is an absurd exaggeration, making a routine procedure look like earth-shattering news. If the public were better informed of what is going on in science, and if newspapers had more contact with everyday research in commercial laboratories, people would not get away with such over-statement. Traditional science has been supplanted by commercial technology, and announcements are usually made in guarded terms that speak of 'commercial sensitivity' and the need to keep quiet about the details.

Research can easily be made to look eye-catching. When scientists first grew rat embryos in culture bottles, the early embryos developed until the cells began to specialize. Developing cardiac muscle cells could be seen contracting. Well, that's what you would expect. By the time the development was being aired, the claim had inflated out of all proportion. The scientists, it was claimed, had observed 'heartbeats'. A current thrust of research is into genes that could indicate when something is wrong with a plant. The 'smart plants' that result could take the guess work out of horticulture. At Edinburgh University the focus of attention is a gene that could signal when potatoes need watering and, at the Institute of Arable Crops Research in Rothampstead, they are trying to locate a gene that could signal when a plant is short of nitrogen. The aim is a simple procedure, which would work like this: plants could be inspected by a hand-held ultraviolet lamp, and would fluoresce under the lamp when diseased or metabolically distressed. By the time the reports were in the newspapers, the research looked very different. The press reports were illustrated by pictures of sunflowers that changed colour like a chameleon.

In this modern world of market forces, a discovery is no longer the important question: what matters is the way the scientist sells it. Modern science policy has much in common with selling double glazing. Politicians, who know nothing about science, ultimately hold the purse strings and are no match for the senior scientist with a list of long words and a lust for money. A hundred years ago it was considered obligatory to involve the media in witnessing science news. Pasteur invited the press to witness his demonstrations. Foucault (who invented the pendulum that shows the earth revolving beneath your feet) was a newspaper columnist, as well as a scientist.

What has changed the relationship in modern times is the way science has been trivialized, which is where the exaggeration comes in. Most scientists are on short-term contracts. They cannot buy a house or securely educate children, so family life becomes almost impossible and everything hangs on getting a grant. Grants are given by committees, who are largely ignorant of the scientists' special skills. The research needs to be over-stated if the committee is to be impressed. Commercial laboratories are as much concerned with putting on a brave front as with achieving anything new. Many biotechnology companies become bigger by company acquisi-tion, which has given rise to a new form of enterprise. Individuals start companies and give them a high profile by salesmanship, in the hope of attracting a buyer and becoming an overnight millionaire. The aim is not to create new science, but to sell a slight idea for an unimaginable profit.

As a result, we have an unholy alliance of journalists hungry for a story, and scientists eager to claim more headlines than their rivals. The result is a new era of over-statement, and scientific honesty suffers. Today's system drives scientists to exaggerate in the interests of survival. The drive is to seek expensive solutions to cheap problems. Science used to be about truth. The modern version, reductionist in approach and blinkered in outlook, is preoccupied with self-preservation.

9

What cells reveal about human behaviour

T HE GREATEST BENEFIT of looking at the behaviour of single cells is that this reveals the seat of our human nature. Multicellular organisms (like ourselves) act like the cells of which we are composed. Human culture is rich in resonances of the way single cells behave. Humans fit together stones to construct a home. Look into a stream and you will find caddis flies which do the same thing. Now take a microscope and peer deep into a pond, or among the fronds of seaweed in the ocean, and there you will find single-celled organisms such as amoebae and Tintinnopsis which can do exactly the same. What we are doing resonates with the way single cells behave. All single-celled organisms embody many of the propensities that we might observe in ourselves. As we are communities of single cells ourselves, I am proposing that humans – and all multicellular organisms – manifest the behaviour that is always found in the single cells that comprise their bodies. There is nothing inimitable in humans. Can this even explain our complex structure, and our sophisticated habits? Can they be modelled on simpler organisms? Let us look back at one of the rumen protists we encountered on page 146. This is *Epidinium*, an astonishingly complex microbe. The organism is but one single cell. Look at it closely and you will see that it seems to possess something that looks like a brain. It is called a 'motorium', and in electron micrographs it shows itself to have something in common with a neural organization. The motorium has a small loop that encircles the mouth of the cell. There is a genetic nucleus, in which the genes in their chromosomes reside, but there is also a far larger meganucleus. This takes no part in cell division, but inside it a great deal of coding information is stored, including instructions on how the cell is to form and to react. *Epidinium* has a tiny pore as an anus. It even has this strange segmented structure containing food storage materials, glycogen and the rest, which to the casual observer looks just a little like a vertebral column. Yet, although *Epidinium* is a single-celled organism, in the way that it has begun to sort out its structure it is beginning to look like a higher, multicellular form of life.

When I proposed that humans behave like single cells, I can imagine your incredulous reaction. 'What about ballooning? What about the Great Wall of China?' Floating through the air is little problem to a cell. The humble pollen grain of the conifer *Pinus* develops two buoyant sacs that transport them for many miles. Clouds of these pollen grains can be seen as

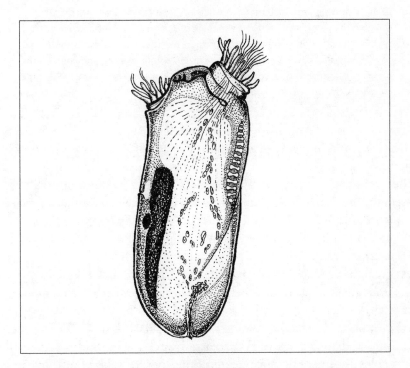

Single-celled life at its most complicated. The single cell of *Epidinium* is an example of the most structurally complex unicellular animal. It has a concentration of sensory and processing protein which is analogous to a brain and even a segmented structure which reminds one of a backbone. In this tiny organism, which lives in the rumen of cattle, we can already perceive some of the traits of large, multicellular creatures.

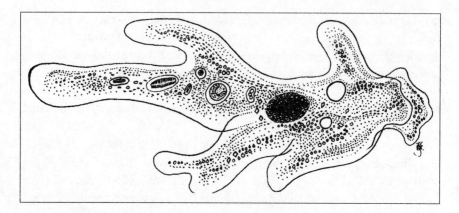

A globule of jelly which feeds and finds its way around. This is a single cell of *Amoeba proteus*, with its dark nucleus clearly visible near the centre. Scattered through the cytoplasm are food vacuoles in various stages of digestion. They contain the algae on which the cell has been feeding. There are also some clear contractile vacuoles. These fill with water and soluble wastes, and come to the surface from time to time to burst and release their contents. These act like the kidneys of the single cell.

a fawn fog over the pine woods in spring. Building with mineral building blocks is found in plenty of microbes. The cell of *Amoeba proteus* may be familiar enough, but other amoebae encase themselves in stony armour. The architecture is perfectly designed to offer maximum structural strength, and the diligent manner in which each shell is constructed makes it possible for us to identify the species of *Amoeba* by the characteristics of its carefully constructed home. One of the best known of these testate amoebae is *Difflugia*. Here we have an amoeba with longer, narrower and less granular pseudopodia than is the case for *A. proteus*. It lives in ponds, and finds small sand grains which it picks up and cements together to form its microscopic home. This is a considerable accomplishment for a shape-less protozoon. Humans discovered how to build with stones a few millennia ago, and we regard ourselves as advanced to have discovered the principles of cementing separate items together to make a protective wall. Living in sea water we find *Tintinnopsis* which uses its cilia to swim about. *Tintinnopsis* gathers fragments of rock, and tiny glassy particles of quartz, and cements these together to make a protective chamber that is shaped like a bell (hence the generic name of the organism). It emerges from its home, holding itself secure with fine, translucent, contractile fibrils.

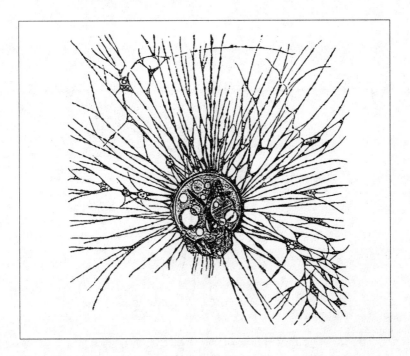

An amoeboid cell with a network of snares. One of the amoebae, *Gromia*, produces an oval shell from which it extends an extensive array of fine pseudopodia.
These stretch far away from the cell, and they trap smaller algae on which the cell
feeds. Currents within the pseudopodia draw the prey slowly towards the centre,
where the cell engulfs and digests them.

Then, when danger comes along it jerks back inside the refuge that it so
perfectly fits.

If you wish to see resonances between these single-celled forms of life
and higher organisms, then, whether it be tortoises or hermit crabs, you
would not have to look very far. Our view of an amoeba is simple: we imagine little more than a shapeless single-celled protozoon, and believe that it
represents the simplest form of life known to science. The way that we are
taught about it, the cell can be any shape you wish. All one needs to do is
draw a wavy line and add a dot for the nucleus, they say. Yet none of this is
true. *Amoeba proteus* is actually distinguishable from other amoebae
because of something about the dimensions, the shape, the outline and the
orientation of the patterns of its pseudopodia. Its shape is actually very distinctive. It even has a head and a tail, although you might not think so. Let
an amoeba crawl along a groove in a glass plate, a cul-de-sac, and at the far

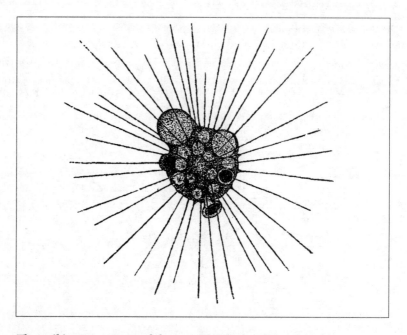

The striking appearance of the sun animalcule. This remarkable organism, *Actinophrys sol*, is also familiar in pond samples. Like *Gromia*, it produces fine projections. These radiate like spokes from a hub. This engraving (like the figure on page 187) was published in the nineteenth-century magazine *Science for All*. In that era, popular knowledge of science was more prevalent than today. This popular publication appeared in Britain, France, Australia and the USA, and enjoyed a wide circulation.

end it has to stop. Then, as we can observe from the movement of the granules inside it, the cell turns round and comes out head first. You have been raised with the belief that an amoeba is so very simple. Simple? An amoeba can do many of the things that humans can do, in terms of metabolism and activity, sensation and excretion, feeding and motility. All these various attributes that we have, an amoeba has too; but amoebae can also carry out a number of actions that humans cannot do. They can regulate their rate of reproduction exactly to match the available food supply and, building a protective capsule around itself, so that when the environment that it needs to sustain life disappears, the amoeba can survive inside this cyst until favourable environmental conditions return. This is quite a neat trick if you can do it. And, of course, an amoeba does all this within a single cell.

Demonstrating the way humans move and walk about isn't difficult.

Two sticks of firewood and an elastic band are quite sufficient to explain to a child the workings of a mammalian limb. An amoeba is composed of water-soluble constituents and ought to dissolve away into the surrounding water. It takes in food in small vacuoles, it excretes wastes through pores that open in the cell surface, and still manages to survive intact. And when it heads off in its chosen direction, it does so without any limbs, without any skeleton, and without the benefit of muscles. Do not believe that it is as simple as they say. If we see the cell as having a whole range of potentiality, then we can try to marry those two opposing concepts of the newly fertilized zygote: it exists as the ultimate microdot, a seed for new life, and yet as the cell from which all later specialized cells can find expression. Multicellular organisms exist by potentiating specific functions in identifiable cell populations, while repressing those inappropriate to that location in the body. Differentiation is the consequence the development, in a pluripotential cell, of only those features that are appropriate for its function in the adult body.

The concept of resonance between cell and organism may throw light on an enduring problem in biology, the evolution of eyes. This has preoccupied writers for over a century, and indeed perplexed Charles Darwin (1809–82). The reason for the puzzle is the tremendous variety of eyes in the animal world. There are 50 different types of eye altogether. Some are compound, others simple; some have a mucilaginous structure, others are hard; some have lenses made from the body surface, others have lenses that are soft and can be focused; some are solid, others are hollow. All fit a basic pattern, in the sense that there is a focusing system at the front and a light receptor at the back. The problem is that different groups of animals have evolved eyes that are constructed in such bewilderingly different ways. If an eye had evolved once, it would be sensible to assume that nature would have repeated much the same design in all the differing life forms. According to the standard teachings of evolution and development, this is exactly what one would predict. The facts of nature contradict that prediction. My concept of resonance could offer the explanation we need. There are eyes within cells. Some of the smallest green algae have perfectly formed eyes. As they are small, we always call them eye spots; but they deserve better than that. A spot is a shapeless dot. These eyes are structurally complex. The typical eye of a microbe has a retina which is usually cup shaped. An image is projected onto the retina by a refractive lens. Running away from the retina are fibrils within the cell which conduct

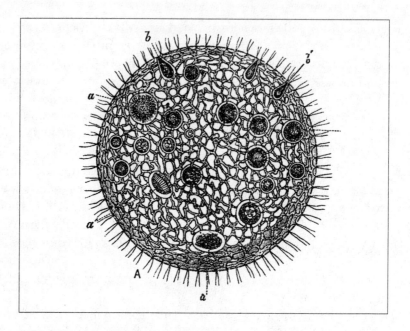

Communal algae which use their eyes. In the pond alga *Volvox* and its allies,
green algal cells grow together to form beautiful hollow spheres. Each flagellated
cell swims in co-ordination with its neighbours, turning the colony like a rotating
globe. Each of the separate cells has a tiny eye spot containing a cup of red photo-
sensitive pigment. Although we cannot imagine how these visual signals are inter-
preted, the colony of *Volvox* is able in some sense to see where it is heading.

the stimulus away. These features are all found within the space of a single
cell, yet these are the salient features of the developed eyes seen in all multicellular animals. The many-celled species are rich in resonances of single
cells, and this explains why all eyes are built on the same basic pattern. The
fact that the different eyes are based on a bewildering array of designs is
easier to comprehend, because we no longer have to seek separate mechanisms that would be necessary had they evolved independently. Now we
can conclude that the construction of the eye is irrelevant to our understanding: what matters is the principle of the eye, and that has already been
laid down within the cell. The way in which disparate animals develop their
organs of sight depends upon the way their bodies are constructed. What
unites them is their dependence upon the resonances of the visual
organelles inside living cells.

If cells can specialize from a multi-purpose original, we need examples

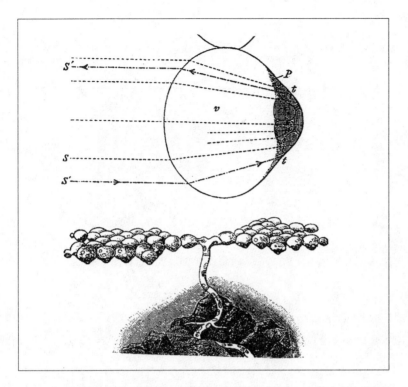

A woodland moss signals the origins of a lens. The Victorian diagram above was drawn by Eduard Strasburger. It looks at once like a section of an eye – but this is a specialized cell of the woodland moss *Schistostega*. It forms raised mats of cells which reflect light like a road traffic sign and seem brightly luminous against dark woodland soil. The phenomenon is called elfin gold. Each cell is specially adapted to concentrate low levels of light.

where we can watch identical mature cells change in structure as they adopt a specific function. Nature provides many examples which enable us to watch what happens. As an example I will choose *Dictyostelium*, an amoeba that you might find in woodland. Were you to find a single cell of this organism under a microscope while examining a specimen of woodland soil, you might well conclude that it was an ordinary amoeba, and leave it at that. But *Dictyostelium* has several phases to a complex life cycle. These little cells exist in woodland among humus and decaying leaves, where they graze on smaller microbes which are part of the recycling systems of the forest. They feed, grow and divide in half, as amoebae always do. When the time comes to reproduce sexually, all this changes.

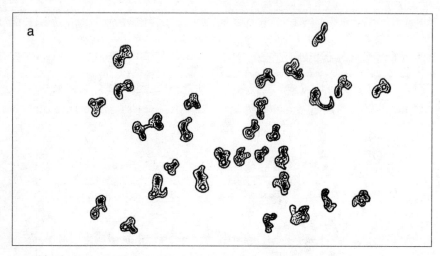

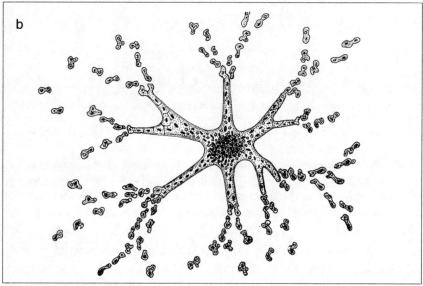

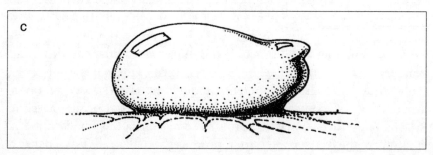

Somewhere in the middle of this great diverse community of separate and free-living cells lies an individual that starts to send out a signal. A time-lapse film of a community of these cells under the microscope enables you to see the signal radiating outwards. Can you picture documentary film of bombs being dropped from an aircraft at war? You see radiating concentric shock waves which spread out at speed from the point of impact. This is much the same as you see in these micro-scopic images; a microbial Mexican wave spreads out from a cell somewhere in the group. It is sending out a signal. It is clearly an outwardly diffusing, behaviourally active chemical, for what happens is that all the cells in that particular community are attracted towards the common focus. As they crawl together they fuse to form a massive strand of cells which becomes drawn in towards the middle. When the organisms have congre-gated together, the mass of cells forms a single body which – as the last few stragglers come in – rises up from the substrate and forms a small globule of life about the size of a match head.

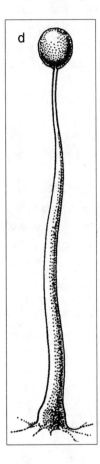

These single-celled organisms have now become a multicellular creature. And then it starts to crawl. If you found one under your microscope you would see it as a multicellular slug-like organism, and would never asso-ciate it with the free-living amoeboid cells that you had seen earlier. Indeed, in many of these slime moulds, the fruiting stage was given one scientific name, whereas the free-living separate cells were given another name altogether. It was many decades before diligent study revealed that they were different stages of the same life cycle. When this slug-like creature has reached a suitable

Opposite and above: **When separate cells come together as one.** Individual amoe-boid cells of the slime moulds exist as separate cells ('a') until the time comes to form a sporangium. Signals sent out by a cell near the centre draw all the cells together ('b') where they congregate to form a single slug-like organism which crawls off to find a suitable place for reproduction ('c'). The cells crawl one above the other ('d') to form a raised sporangium. From this, spores are released to perpetuate the species.

place, it stops moving. The organisms change their co-ordinated behaviour once again, and the rounded body begins to rear up into the air, forming a tower in which the amoeboid cells are climbing up over each other. At the very top they spread out to form a towering structure several millimetres tall, like a microscopic space-age spherical city. This silvery, moist structure is like a tiny pin. After some complex reproductive processes that are not yet fully understood, cells within that globular wall form hardened walls and become spores. Then the sporangium breaks and these spores blow away to colonize new areas of the woodland on which they chance to land. This is an extraordinary life cycle. It is important to note that these amoebae do not make the mistake of joining together with members of other strains or disparate species. They have the ability to reject all the other signals from microbes that abound in the woodland soil, and at the end of the day the cells with which they choose to team up are recognized as different, and desirable, out of all the others that they encounter.

Just as the shell-building proclivity of protists is reminiscent of comparable patterns of behaviour in many other walks of life, so the coming together of separate cells to form a single body is found in other groups of microbes. There are bacteria that join together into a single body in exactly this way. The bacteria of the genus *Chondromyces* also send out similar signals. When the right signal is detected by the right bacteria, they begin to migrate towards each other to form a clump of cells. This unified body grows up above the surface of the substrate and produces sporangia at the ends of branches sent out from the central pillar, from which dry spores are released.

Equally intriguing are the motile colonies of bacteria, which have been known about for many decades but about which little is yet known. A single rounded colony of these bacteria may contain billions of separate cells in the space occupied by a pin head on the surface of an agar plate. The entire colony crawls. They are composed of separate bacteria, not aligned in any form of matrix, but growing together and moving together so that the entire colony has its own 'sense of direction'. Occasional organisms are left behind as the parent colony crawls across the plate, so you may see where it went as daughter colonies start to form in the wake of the moving mass. The complete, match-head-sized community of disparate organisms is moving on its own. Some of these organisms are used by researchers to model simple cells, and *Dictyostelium* is popular in cancer laboratories. I think that they can tell us far more about the way

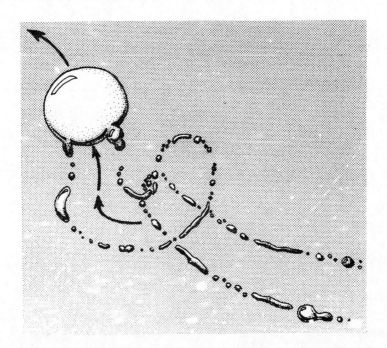

Bacteria that can form a single colony. Some species of bacteria form colonies which move as one. This colony, grown on an agar plate in the laboratory, moves of its own accord like a single organism. From time to time, groups of cells break free and form their own tiny colonies. These colonies also act like primitive organisms and can travel in concert across the surface.

all organisms relate to each other, and how they tell each other apart.

We ourselves are teeming with amoebae. That is not what we call them, but that is what they are. These are the white blood cells. They were once called phagocytes, after they were investigated by the Russian scientist Ilja Metchnikoff (1845–1916). These cells consume bacteria and other foreign bodies in the blood, and were first observed doing so in Germany by Ernst Haeckel (1834–1919). These cells have granular cytoplasm and a lobed nucleus that looks as though it is made up of several distinct compartments. The use of the word phagocyte dropped out of favour between the two World Wars, and they became known instead – because of their appearance – as polymorphonuclear granulocytes. Science likes to translate simple words into complex terms by multiplying the number of syllables, usually by three. In this case the change from phagocyte to polymorphonuclear granulocyte goes up from three syllables to eleven,

Independent bacteria form a single living structure. Some bacteria behave like the slime moulds. Their normal life is spent as independent microbes, but this changes at the sexual reproduction phase, when spores are produced. The separate cells congregate together, producing a single raised organism which reaches up into the air. Here we see the beginning of separate organisms combining to form one: the origin of multicellular life.

somewhat above the average. This is how the language of sciences keeps outsiders in their place; in today's laboratory they are only called polymorphonuclear granulocytes in a formal environment or when a visitor comes to the laboratory. In research they are just called 'white cells'. In fact there are many different types of granulocytes, including basophils and eosinophils which are named after their affinity for microscopic stains. There are other white cells too, including the monocytes which are the largest of them all and the lymphocytes (concerned with immunity) which include the smallest.

What unites them all is their free-wheeling, independent way of behaving. These cells seek out invaders and foreign bodies and do all in their power to destroy them. Lymphocytes are trained to recognize friend from foe in the thymus gland, and there are other systems within our bodies that keep the white cells trained to repel outsiders. There is no way that you can control these cells, although they are part of your body. They are self-propelled, self-motivated, entirely self-contained. No hormone influences where they go, and no nerve communicates messages to them. These cells have their own decision-making processes and are entirely individual entities. Yet they are among the cells that make up the human body. There could be no greater reminder of the way in which we are nothing more than colonies of separate cells, which function together to produce a person. If the cells detect foreigners, then they instinctively seek to extirpate them. They never run and hide, but devote themselves to destroying a foreigner even if they are themselves sacrificed in the process. The urge to destroy is overriding, even if the foreign body is there for the good of the whole community. A transplanted organ, for instance, is attacked by the host's white cells unless this immune response can be dampened down. Sometimes we do this by the use of drugs; we can also cloak the cells of a donated organ with genetic markers which will make it seem like 'self' to the recipient's cells. In essence, we could be dressing the donor cells in a disguise that the recipient's body will identify as friend, not foe.

At Cambridge, England, high hopes were raised by the possibility of creating transgenic pigs that could be tailored for a patient in need of an organ transplant. The pigs would carry marker genes from their patient, so that an organ transplanted from the pig to the human recipient would be recognized as 'self' by the new host. It was claimed that this would eliminate the problems of organ rejection. Objections have been raised because of the chance that pigs may harbour viruses to which they are accustomed, but which could trigger a disease in human beings. This would be an extraordinary consequence of a well-meant act: perhaps we could unleash an epidemic of a virus disease that is trapped within the genes of a pig through transplantation.

The evidence that we have does not support these unsettling possibilities. During 1990–93, ten diabetic patients were given replacement cells of the islets of Langerhans from pig pancreas. These islets are the tiny glands within the pancreas that secret insulin, and their replacement was hoped to help cure the patients of their diabetes. Interestingly, several of these patients developed antibodies to pig viruses; this implies that there are indeed latent viruses in the pig tissues, to which the human cells reacted. However, nobody became ill. We need to be absolutely certain of what we are doing before we start tinkering with transgenic transplants. We also need to beware of scare stories without foundation. There remains a possibility that, in dampening down the body's tendency to attack an invading cell line, we are creating unforeseen difficulties for the future.

The complexity of the cell's ability to attack the outsider has resonances throughout the animal world. Nature is rich with on-going territorial disputes and civil wars in human societies bedevil us as much today as in previous centuries. People seem to take a perverse delight in attacking each other and we now seem to accept that tribal conflict and intercultural strife are normal. There are few theories that explain why we seem to be endlessly at war. The explanations that have been laid down to explain why people conflict with each other do not always fit observable facts. Military theorists say that territorial aggrandisement or economic advantage is the cause of warfare, but the most cruel acts of war, where there is necessarily any territorial advantage, are not. What we see are neighbours, people of a similar culture, attacking each other in acts of deliberate defilement. Nobody wishes to claim new land; what is attained is the destruction of a way of life and the disfigurement of the vanquished. This is the machination of instinctual hatred of one form of human culture for another. There are many areas in which rejection of this sort – the animus of hatred – exists within communities of living organisms. Let us try to synthesize these ideas and see how a deep-seated rejection of non-self can explain human strife, and even provide an engine for evolution. Evolutionary theory is one area in which strong forces of what you might call 'selection against', rather than 'selection for', seem to operate. This may provide the elements missing from traditional theories of evolution. Charles Darwin's position is widely misunderstood. He has been internationally heralded as the progenitor of the fashionable view, and the need to be a 'Darwinian' is celebrated almost as though it were some kind of religion. To some modern scientists, that's exactly what it is. 'Darwinism' is now a tenet of faith. 'Darwinian evolution' is at the basis of modern biological philosophy, and those who doubt this view are regarded by their fellows as heretics or infidels. Darwin was a great a popularizer of evolution, but he was not its originator. The word 'evolution' does not even occur in the text of the original *Origin of Species* until the sixth edition. The presentation read to the Linnean Society was not by Darwin alone, but was a joint paper with

Alfred Russel Wallace, whose *On the Tendency of Varieties to Depart Indefinitely from the Original Type* was crucial to the understanding of evolution. At the close of that year, the Presidential report on the activities at the Linnean Society did not (as you might have expected) hail the paper as a major landmark. Instead, he noted dourly that nothing of any great importance had happened during the year.

There is a school of thought that insists that the theory of evolution rightly belongs to Wallace, rather than Darwin. The truth is even worse: by the time the Darwin-Wallace paper was read, the publication of natural selection was at least some 30 years old. The theory was expounded in an earlier book on arboriculture, published in 1831, which became well known to Darwin. Its author was Patrick Matthew (1790–1874), a Scottish fruit breeder. The book, *Naval Timber and Arboriculture*, included an appendix on natural selection. He said:

There is a law universal in nature, tending to render every reproductive being the best possibly suited to its condition... its kind appears intended to model the physical and mental or instinctive powers, to their highest perfection.

Matthew went on:

Nature, in all her modifications of life, has a power of increase far beyond what is needed to supply the place of what falls... those individuals who possess not the requisite strength, swiftness, hardihood, or cunning, fall prematurely without reproducing... their place being occupied by the more perfect of their own kind.

Charles Darwin was to concede that:

I freely acknowledge that Mr Matthew has anticipated by many years the explana-tion which I have offered on the origin of species under the name of natural selec-tion.

He concluded:

If another edition of my book is called for, I will insert a notice to the foregoing effect.

Clearly, the concept was a growing facet of contemporary awareness before Darwin's book was written. Later he changed his mind, and wrote conde-scendingly that:

An obscure writer on forest trees clearly anticipated my views ... though no single person ever noticed the scattered passages in his book.

Since then, few have ever heard of Matthew, though he must have had an effect on contemporary society because his book was banned from libraries in Perthshire. In espousing natural selection decades before Darwin, he reveals that the idea was beginning to grow in the attitudes of the time. His writings are rich in resonances of contemporaneous cultural conventions, as when he clearly reflected the view that it was fitting for the British Empire to dominate the world. Some of Matthew's attitudes have the look of fascism, as when – in the *Emigration Fields of 1839* – he supported British colonials cutting down a few aborigines as you would 'the encumbering trees'. Matthew retreated from the racist view of his middle years, and wrote a letter to Darwin in 1862 which showed the extent of his retrenchment: 'I am not satisfied with my existence here to devour and trample on my fellow creature.'

This publication of natural selection so long before the *Origin of Species* makes Darwinism seem a relative late-comer. I believe that Darwin was reflecting a burgeoning scientific consensus, and the existence in the popular consciousness of the principles of natural selection underpinned the commercial success of his book. Today's Darwinists worship his marvellous synthesis; indeed, there was a solid weekend of programmes on Charles Darwin on BBC television in the spring of 1998 claiming to show that his intellectual breakthrough and his coinage of evolution influenced everything from politics to architecture. The theory was claimed as the greatest single advancement in philosophy for centuries, and the notion of humankind as descended from the apes was presented as a stunning revelation. I am suggesting the opposite – that Darwin was reflecting an existing concept. I can even quote his own words to support this apparent heresy, because, in the introduction to *The Descent of Man*, he wrote:

The conclusion that man is the co-descendant with other species of ancient, lower, and extinct forms is not in any degree new.

In their blind eagerness to hail Charles Darwin as something close to a deity, I never see these words quoted by contemporary Darwinists.

Does this show that the name of Darwin is wrongly associated with natural selection? There is a description even earlier than that of Matthew. It argues clearly that: 'the strongest and most active animal should propagate the species, which should thence become improved.' This is Darwinism in a nutshell, and these are Darwin's own words – though not those of Charles Darwin, but of his grandfather Erasmus, published in his

Zoonomia of 1794. Charles Darwin later admitted that he had read *Zoonomia*, but stated that the 'similar views' in his grandfather's writings did not produce 'any effect on me'. Erasmus Darwin clearly coined the concept of survival of the fittest, and ideas such as this may have under-pinned Matthew's theory, which Charles Darwin also acknowledged. His work actually said very little about the origin of species at all. His work was concerned with the pressures that maintain diversity of species, but speciation is the moment at which the dichotomy between one species, and two subsequent species, becomes manifest.

Suppose, for example, you were to take a group, a community, a field of medium-tailed shrews. Were you to leave them in the field and come back in a million years' time, it could be quite possible that, by then, you would have two discrete species: a short-tailed shrew and a long-tailed shrew. The mathematical model, which has been constructed in order to show how speciation has occurred, has a kind of three-dimensional groove, a divided Y-shaped groove, down which marbles run like marbles. Some go down the left branch, others the right. This is retrospective modelling of what goes on when one species has become two. But what lives with me is the fact that – as the marble is running down the groove – the groove, and the division in the groove that lies in front of the rolling marble's position, do not yet exist. It is the marble itself that is the cursor marking the rate at which evolution takes place. So why, when the marble actually comes down to what we now diagnose as a dichotomy, would the marble start going towards either the left or the right? Once the groove is built it is easy to see how the marbles behave. But the marble creates its own groove-generating process. It is at the forefront of evolution, and can have no sense of knowing where it is going.

Let us return to our shrews as a convenient model, and examine what this means. It is quite easy to assume that long-tailed shrews – that is to say, members of the medium-tailed shrew species, but with slightly longer-than-average tails – might in fact tend to breed more at one end of the field, with the shorter-tailed individuals breeding at the other. This is what Darwin taught, because geographical separation was to him one of the most important distinguishing factors that led to speciation. If you kept species apart, then the individuals unable to adapt to disparate environ-mental experiences would die down while others increased. In this way, given enough time in these two geographically discrete regions, you would have two separate species. In the real world geographical separation rarely accounts for this phenomenon. Many species (butterflies are a prime example) have evolved in diverse directions while coexisting. So we now have to address why it is that a given community of organisms – like our medium-tailed shrews – would ever become two species. You might deduce: 'Well, the shrews with slightly longer tails breed with each other, and so do those with slightly shorter tails, and so at the end of a million years we have two discrete species.' This works only if you bring in a new concept: not merely that the longer-tailed shrews are breeding with each other, but that these longer-tailed shrews don't like breeding with the shorter-tailed shrews. It is not merely a question of organisms wishing to interbreed in groups, but of not wishing to interbreed with other groups. There is a principle in evolution of the rejection of the dissimilar by higher organisms. Now, what it might be we need to debate, and why it might be we need to discuss.

The theories of evolution omit one crucial aspect of sexual selection. The occurrence of female choice of a mate is well documented, and to Charles Darwin and others it has always been an engine of evolution. One of the principal manifestations is the mating rituals which we see through-out the animal world. Courtship displays, particularly of birds, are some-times bizarre. At the microscopic level, we see them in protists who circle each other as they decide whether a potential mate is suitable. We even indulge in mating rituals ourselves, involving the painting of the lips and eyelids, the adornment of the body in clothing and the anointing of the skin with perfume. As we indulge in them, we naturally accept them as normal, though they are bizarre when considered objectively. If mating rituals did not exist, nobody would have perceived a need to invent them. They are now seen as a keystone of evolutionary progress, and the notion of female choice is, currently, of growing interest. Female choice is, however, misunderstood. We think that it means that a female makes a positive discrimination in favour of her chosen partner, but this cannot be the key. She is only in a position to use this positive criterion if she has already made many negative assessments. The choice of a mate can only eventuate if other (earlier) would-be suitors have been rejected.

If one female is offered nine suitors, it is not enough to say that they all displayed until she found one that proved to be acceptable. That would only work if a male were to come along, display and then – if nothing happened – depart. But in fact, before she can make her final choice, a female must decide in some way that she does not want other available suitors. Here is a clear example of the need for the rejection of the unfit suitor. An obvious objection to this proposal is that, in many species, successive males mate with a given female, whether she likes it or not. The work of William Eberhard in the USA is showing that some of the mechanisms of control are concealed within the female's body. In many species, the sperm from successive males are stored within the female, but she then still makes internal choices as to which sperm will fertilize her. She produces cytokines, chemicals that signal to cells how they should behave. Some species have complex and elongated passages through which the cells must pass, and use these to filter out the wanted from the unwanted sperm. The females can detect genes for attractiveness or virility, and can elect to have these sperm fertilize her ova. Thus, female choice really governs evolution and sexual selection. If Eberhard is right, many of the mechanisms are internal, and hidden from male intervention. Yet we have to grasp the full implications of selection. It is not a question of knowing what you like – rejecting what you recognize as alien is even more fundamental, and must take priority.

In our daily lives as humans we recognize dissimilar groups. They are coded by dress, behaviour, by speech and pet phrases. There are few far-right skinheads bent on racial harmony, just as there are few Conservatives wearing open-toed sandals and Oxfam sweaters. It was, of course, a Labour politician who spoke at a formal dinner in London wearing an informal brown suit; a senior local government official has just been dismissed from the office for wearing jeans and a tank-top to work. Clothes are one form of marker adopted by groups of like-minded persons, and they set boundaries of style which cannot easily be changed. The student body, all wearing jeans, will be quick to retort that 'we don't want to wear suits'; yet denim is a uniform. Jeans are the businessmen's suit of the young.

Our tendency to bunch together in groups, gangs or tribes is the puzzle. In my view the main feature is not the bringing together of like-minded persons, but the rejection of non-self. The choice of clothing, and the words that are used, are not merely there in order to attract similar people: they are there to express rejection of dissimilar individuals. A phrase, uttered *sotto voce* through expired breath – 'women drivers!', 'skinheads!', 'students!' – is a reflection of the derision that betokens this rejection. The wish to be distinctive is reflected in speech as much as in clothes. One of the enduring mysteries is why languages and accents have become so disparate.

This is because language has two purposes. Its use as a means of communication is only part of the value of language. It is even more useful as a group indicator, a mark of self or non-self. If language were a means of communication between peoples, there would be a tendency for languages to merge. Instead, they became increasingly dissimilar as the centuries unwound. We have seen this in our own time. For a decade or more, in the 1980s, the yuppies in London started to say: 'Okay, yah?' as a kind of absent-minded punctuation. At the same time, youngsters in Los Angeles started to intone vowels with a fake English accent: 'Hellow,' they would say to greet each other (and to annoy their parents). Here we are watching language developing – only members of the group spoke thus. Scientists use special terms, and we believe that this also is to aid communication. In fact it is quite the contrary: the scientific terms are meant to hinder understanding by outsiders. As I have often remarked, in haematology laboratories an erythrocyte is known as a 'red cell'. They are routinely called red cells. However, when an outsider comes into the laboratory, they become 'erythrocytes' once more, and they remain erythrocytes until the stranger has gone. A key purpose of language is to mark out a group as privy to one specific type of communication.

Treasured by those who speak them, unfamiliar forms of speech can be fiercely resented by ruling groups of a different cultural persuasion. Languages develop in different directions to keep communication within the group, and not to let it spread. We can see how much this is resented by the way that alien languages are extirpated by a ruling group. In the time of General Franco, Catalan was banned in Spain. Publications in that honourable language were illegal, and so was the possession of the Catalan corpus of literature. In the nineteenth century, the use of Welsh by Welsh children in Wales was punished by teachers versed in the culture of England. Any child heard using the ancient Welsh language was given a wooden token to carry, called a 'Welsh not'. That child would hand it on to any other child whom they heard speaking Welsh, and so on until the end of the day when the pupil still in possession of the token was punished. Languages are a cultural indicator, a marker like one on a strange cell in the body. It is significant that the banning of a language is regarded as so important in cultural disputes.

This explains why the current debate about a United States of Europe is a non-starter. There are huge cultural differences between neighbouring towns – between Nottingham and Derby, Brighton and Hove, New York and Jersey City, Los Angeles and San Francisco, Tokyo and Osaka. The USA has remarkable cultural cohesion, even though its many accents and the different ways of life vary from state to state. This is because the only true Americans are red-skinned people largely confined to reservations; the white Americans we see in the media are rooted in European cultures that have blended to produce a new and all-embracing understanding. Europe is not like that. Europeans have spent thousands of years developing different cultures, rather than creating a new cultural convention that unifies a newly emerging people. Shades of behaviour give a tremendous vitality to the peoples across that historic swathe of land, but forcing uniformity upon them poses great problems of which the bureaucrats have little understanding. We need to celebrate the cultural richness and diversity of Europe, rather than acting as though Europeans are all the same. I have served as President of a European writers' organization based in Brussels, and indeed have worked with the Commission. For me, the thing that marks out the European decision-makers is that they have created a great culture of their own. It is, to be unkind about it, the culture of the inward-looking, narrow-minded, gin-and-tonic-quaffing bureaucrats. Wherever they congregate they always attract others like them. Although they rule Europe, they adopt a common stance and a similar way of dressing, largely heedless of the cultural diversity that you find in Europe. We believe that, in Spain, the people speak Spanish. Not true: they speak Castilian, Catalan, Gallician or Basque, and there are divisional sub-languages within these. You believe that in Germany everybody speaks German. Not so. In the Saar there is a language, a little like the Luxembourgois spoken to the north; there are even the Wendish people near Berlin whose tongue is unrelated to German. The language of Provence is hard to understand if you are up in Paris, and Breton survives in north-western France. There are at least six languages native to Britain (English, Welsh, Gaelic, Manx, Norman French in the Channel Islands, and Cornish, now in the ascendant). All across Europe there are cultures within countries, and complex cultural differences that others find hard to understand. Many of them relate to an ancient way of life. Accents often change as does the landscape, and sometimes a harsher landscape is matched by a harsher way of speaking.

The whole basis on which the European Union has operated is founded upon a myth like the Emperor's new clothes. When the bureaucrats began to flex their muscles in Brussels, they set out to bring uniformity to Europe.

They actually had proposals for Euro-jam and the Euro-sausage. There were going to be specific amounts of ingredients, defined by law. Now, the sausages eaten in Cambridgeshire and Lincolnshire contain sage. You don't normally expect to buy them elsewhere in Britain. The sausages they eat in Bavaria are quite distinct from those they enjoy up in Frankfurt. Those people in Brussels more recently came out with the order that Stilton cheese may maintain its name, because it is associated with the little village of Stilton. But Cheddar is to be freed from control as a name for cheese, because it is too well known and international to be truly associated with the Somerset village of Cheddar! Yorkshire pudding can be made in any country whatever, and indeed is more often made in areas other than Yorkshire; whereas Champagne is a name you would be sued for using to describe a wine produced anywhere at all outside the narrow confines of that specified region of France.

It is a bureaucratic nightmare, run by self-serving politicos who love their Mercedes, adore their unlimited telephone calls and web access, and who are never happier than when they are arbitrating over other people's affairs. Subsidiarity was brought in to engender the notion that some decisions might be taken by cultural units within Europe. It ignores the extent to which people have different beliefs (and I will come to an example of that in a moment). But the term shows what a confidence trick subsidiarity truly is. The notion of being a subsidiary implies that you are under the control of something bigger. In other words, it would be Europe that is allowing you to have your differences. It should be the other way round. The peoples of Europe should be insisting that we are already different, and that a much reduced Brussels executive should concern itself by solving some of the real problems we face. If it means doing away with passports, and doing away with monetary restrictions and trading control, then so be it. These are the inventions of governments. There is no real reason why you shouldn't travel around the world free of bureaucratic restriction if you want to. Far better is the Swedish notion of *allemansrätt* ('everyman's right'), a concept of freedom of movement. It is not regulated by law, but by custom, and prevents landowners from keeping people off their land, with the exception of the area immediately surrounding their dwelling. Compare that with the norm in England or America – there you would be challenged by the landowner if you stepped off the designated pathway. It is behavioural markers like these that set out the way we live.

Behavioural differences can cause offence. Let me give a specific example: I present to you three dining tables. One is in Boston, Massachusetts; the next is in Cambridge, England; and the third is in Uppsala, Sweden. In Cambridge, England, if you laid a formal dinner table and you stood upon it cans of beer, people would think that was extremely infra dig. Beer should be poured into a good crystal goblet or – even better – a fine pewter tankard, to be quaffed in a civilized fashion. If a carton of milk were standing on that table, it would also be considered totally inadmissible. Milk is poured into a bone china jug, and then dispensed properly (before the tea, but after the coffee, in the traditional manner). Now let us move over to Boston, Massachusetts. In that city, if you pour out the beer into a tankard or a glass you mark yourself as wimpish and beyond the pale. Everybody in America drinks beer straight from the can. It is the macho way to do it. Admittedly, the people that I visit in America tend to cut down on the habit after I point out to them the likelihood of dogs pee-ing onto grocery cans standing in the store, and the crusted effluvium that accumulates from people coughing as they pass. After that revelation they do tend to revert to pouring the beer into a glass. And, what's more, they start wiping the can first, before they transfer the contents. But – just try putting a carton of milk on the table in Boston, and that would be disapproved of... 'This is a formal dinner party! Take that carton of milk away and pour it into a pitcher!'

Now we fly across the world to Uppsala, Sweden. The Swedes, like the other Scandinavian peoples, delight in consuming the many forms of milks and fermented products that have developed over the centuries. One of them is the string-milk, in which those colonies of *Streptococcus* have grown. The milk comes from the carton and looks like phlegm: white milky phlegm, but phlegm none the less. There are all sorts of other glutinous yoghourt milk products, and so in Sweden you would always find the milk carton. It is not just put on the table, but reverentially placed there in a position of honour. However, drinking beer from the can at a dinner table in Sweden would be considered most impolite. It is considered an uncivi-lized and insanitary habit. How interesting it is that the behavioural norm in Britain would be to have no cans or cartons on the table at all; in Sweden they would say that there is nothing wrong with milk from a carton, as it is thoroughly natural – but get rid of the can. Yet in Boston you would be told that a can of beer is the most civilized thing in the world, but you should not see a milk carton on a formally laid table.

These behavioural nuances are indicators of conventional, acceptable behaviour. Even slight departures from the norm can cause irritation in someone used to a different way of behaving. There are countries in which you keep the same plate from one course to the next, but expect new cutlery; states in which you keep your cutlery, but change the plates; cultures in which both are changed as the meal progresses. A failure to observe the norm does not merely mark one out as an outsider, but causes an inborn disquiet. Cultures really do keep distinct by rejecting patterns of behaviour that they do not espouse.

This is why primitive tribes traditionally waged wars between groups who, to us as outsiders, would have seemed quite similar. They did not merely wage war in the sense of wanting to capture their land, because they usually took trophies and then went back home when they had finished. And the purpose of the conquering was not merely to conquer land, but to subjugate people, to rape, to disfigure, to mutilate, to cannibalize and to take home the liver or the scalp. It was Konrad Lorenz who coined the concept of humankind as a murdering ape. He studied the way in which geese could be induced to follow him around. Newly hatched geese tend to identify the visual appearance of whatever figure they see shortly after they have hatched. There is a critical phase for this imprinting phenomenon, as we now call it, and during that time the gosling has mechanisms for the identification of 'mother', ready to be wired or programmed. This is true of species other than geese – it is the reason why, in the nursery rhyme, when Mary had a little lamb, it followed her to school. We must examine the nature of this learning, at its instinctual level, and pose the idea of genetic potentiality that remains potential until the experiential programming of that potential is realized.

Some work just published in Ohio has shown how honey bees show an extraordinary affinity for their siblings. If you mix up a number of worker bees from different communities, bee communities in different hives, they very soon sort themselves out into sibs. A person used to typing insects by sequencing their genes would find that level of distinction very difficult to undertake. Yet the bees manage it perfectly well. They do not select their own kind merely by positive taxis, by homing in on their kin, because that must be a secondary effect. What must come first is the rejection, or the recognizing of non-kin, of the earlier workers that a particular bee has come across. A bee only remains available for the partner it is eventually going to choose, or the group it is eventually going to join, by cognitively rejecting the alien bees that it meets earlier. Here is another example of the innate propensities for rejection that exist.

Other research at Cardiff University by Michael Claridge has shown how Indonesian leaf hoppers tell ostensibly identical organisms apart. In the hand, under the lens, the two species that they have studied are morphologically identical. It is impossible to tell them apart. Using scanning electron micrographs, if you look very closely, you might possibly tell one species from the other. But even to the gifted and experienced human observer, distinguishing these two species is pretty well impossible. The two species are found in different regions of Indonesia, and when Darwin wrote of such phenomena he concluded that the reason such species did not interbreed is because they were distinct, and because their territory hardly ever overlapped. But since the evolution of agriculture has brought about high-density farming in Indonesia, the geographical zones in which these two species exist now overlap. The two species of morphologically identical leaf hoppers coexist in the same vegetation in Indonesia, but they don't interbreed. The reason that you can tell that they don't cross-breed in nature is quite simple. If you take them and confine them in little jars in the laboratory, then – probably out of boredom and because of the confinement in proximity – they will subsequently interbreed. They do not do this often, but sometimes it occurs. The question then arises, how do we tell them apart, and how can we tell that they are different species? The distinction lies in their mating calls. If you obtain a voice print of the mating call of these leaf hoppers, then one species has its vocal activity at one end of the scale, and the other has its vocalization at the other. They stridulate on blades of grass at different frequencies and with a different timing in order to distinguish one from another. Leaf hoppers elsewhere down the blade of grass, who feel this vibration transmitted along the blade, quickly recognize whether this is another of their own species. If the alien sound signal is transmitted down the blade, then they realize that this is the other species, and they reject the chance to mate. As they can be encouraged to interbreed in close confinement, it can be shown that the hybrids have their voice spectrum in between the others. It lies in the centre of the graph. So, by collecting hoppers in the wild and putting them into a small sound-measuring chamber, it is very easy to distinguish between species A, species B and a hybrid. The hybrids have never been found in nature. Here too we must have an example of my postulated mechanism for rejection. The leaf hopper is sitting on its leaf, longing for a mate, when the wrong message comes down that blade of grass. So, even in these ready-to-mate communities, there is a clear indication that they reject signals sent out from an alien culture.

There is much importance in the study of learned responses, coded as criteria of selection mechanisms. These criteria, which enable us to select things that we like, must also (and more importantly) code for things that we reject. A paper in *Nature*, for example, has discussed the phenomenon of spur length in pheasants. In this case it is clear that spur length in the male pheasant is plainly the signal that is recognized by the female of the species as an indication of whether she wishes to mate, or, more importantly, whether she wishes not to mate, with a specific male pheasant. The critical response to this paper argued, first, that the standard interpretation of Darwinian selection predicts that attractive sexual characteristics which enhance mating success should effect male survival, and, second, that if spurs are so beneficial to male pheasants, then why are they not rapidly increasing in length in the males of that species – or, I might add, the males of all other species, too? Why is this decoration lacking in females? It was argued that there was no additive genetic variance for long spurs. But a simpler and more novel explanation was offered in this paper. It was that spur length is a condition-dependent characteristic. In other words, the young female pheasants have to learn to recognize the target as an important indicator. This, it was concluded, was why the females were selecting for long spurs. It is almost like people with Mohican haircuts recognizing that culturally they have much in common.

These examples fit the predictions of the general theory that I am advancing here. It is that selection is taking place by the rejection of the 'un-approved of' partner. It is always important to remember that a female does not merely positively select a viable mate, but must embody innate mechanisms to reject the unwanted ones. 'Selection for' can only exist in a multiple-choice situation where 'selection against' also exists. In the real world, the separation of encounters is chronological and she must know that in declining to pair off with the first few males that come along, she is saving herself for one that she is going to identify positively as a viable mate. In the past we have homed in on the second phenomenon, but the positive response can only exist if innate qualities of rejection have applied earlier on in that process.

We often neglect the importance of learning. At every stage of life, at each moment of development, we respond to learned input, and are tuned by what we experience. We have complex systems to correct for experiential input that is of little value, although of great power; we lose sight of the way in which we can change through our lives, and can adapt our attitudes as we respond to new revelations. In many cases, this ability – which exists widely throughout the kingdoms of life – is dismissed as a mechanical system. Once learned, never forgotten; once wired, never changed. Life is far more subtle than that. Let us look for a moment at a classic example of mechanistic thinking: the reflex, such as blinking when a blow is aimed at your eye. This is a response to a stimulus, selected by inborn criteria and designed to protect the individual. Such a property is explained by evolutionary theory, and it involves an inborn reflex. Reflexes have physiological roots, which might well be a response to a change in circumstances – such as the 'fight or flight' response, involving the raising of pulse rate, and a boosting of the circulation during effort, which is also an unconsciously regulated response. But there are other codes that we have to learn to associate with circumstance, and these Pavlov named the 'conditioned reflex'. They exist as circuitry waiting to be experientially wired, if you like. A properly critical investigation of this area really does need to be carried out. Even when a conditioned reflex is fully established, there is always the possibility for re-wiring, or for modifying the wiring. Sexual behaviour in humans is regarded as fixed, but in fact erogenous zones, and erotic behaviour, can be learned over a period of years. People handicapped by the impairment of a set of senses can learn new ones, which can restore one sense through the mediation of another. In the original experiments of Pavlov, dogs were taught to associate the ringing of a bell with food. Then, when the bell was rung and no food was present, the dog would salivate anyway. That's fine: it is an experiment that we were told about in school. But what interested me when I was presented with this in school was this. Suppose it is a different bell? Or what if it is a recording of the same bell, but the frequency is raised – the tape speed is slightly increased? Or suppose that it is an electrical r-r-r-ring instead of a single 'ding'? Suppose just one strike of the same bell was heard? How far can one depart from the conditioned stimulus and still maintain the response? As far as I am aware, no research has been done into this.

We make general distinctions all the time, and rarely consider the basis for a judgement. In our own experience, people may say, 'Oh, I hate ham,' for example. But, although it is also pink, they may like corned beef, or tongue, although it is similar in texture; they might like pork, which really is ham (but without the nitrite preservative) and so, though it is a different colour, it remains the same meat. In other words, it is not good enough to say that a specified response is elicited by a given stimulus. How far would you have to go from that stimulus to find that the wiring still triggered the conditioned response?

This is a large field replete with ignorance, and rich in areas that require investigation. It could show that a conditioned reflex might not truly be a 'reflex' at all, but a learned reaction. It suggests that learned responses to situations might be un-learnable, or even re-learned. And if that is possible, it might be that we can explain how the instinctual rejection in conditions of phobia could be subtly re-programmed. Rather than acquainting arachnophobic people with spiders until they start to like them, it might be better to modify the picture slightly, until you arrived at something like a giant ant which – although morphologically similar – didn't cause the phobic rejection. I think that by mapping out how far we can go from the stimulus that causes rejection, to something similar but different, we can probe a new area of human psychology.

The world of microbial recognition is a fascinating field. We now know a vast amount about how a single cell operates. We understand the biochemistry of a single cell. Yet for all this, we know far less about how two cells function. Understanding the mechanics of a single cell is comparatively easy. As in the model of the blind watch maker, it is a case of inspecting the gear wheels and trying to see how they fit together. Molecular biology is rather like looking at the transistors in a radio and guessing how they work. The important thing is not the radio, but the programmes; not the watch, but the race it is used to time. How a cell works is not the problem – we need to know how two cells function together. They have intricate mechanisms that allow them to identify each other and to tell one strain of cells from another. Cells signal to each other using nitrous oxide (a gas we know better as a propellant for whipped cream aerosols, and as the laughing gas used to ease the pain of childbirth). The latest microscopic techniques allow us to see bursts, puffs and waves of calcium running along cells which act as a form of language. Already we are beginning to identify some of the chemical signals that bacteria use to communicate with each other.

A single swimming cell can suddenly stop swimming and go hunting in search of a mate. *Paramecium* is an example, because this perfectly streamlined ciliate spends its life grazing on bacteria. If a suitable mate comes along and, if the season is right, then both cells start to spiral round each other as they swim. They are inspecting each other, and picking up subtle signals. Often they will break away and continue grazing, but if they approve of their partner they come closer together, fuse, and have sex. Sometimes protists look completely unsuitably designed to have sex, but still they manage to do it. *Loxophyllum* is a flat and tenuous freshwater organism. It looks like a floating flame, less than a tenth of a millimetre from end to end. It is covered by beating cilia. *Loxophyllum* is translucent and exceedingly thin, like nothing more than a piece of protoplasmic membrane. It shows food vacuoles and a very complicated nuclear

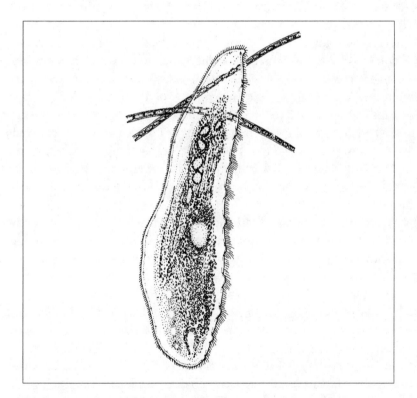

A thin sheet of life, a single cell with an active sex life. This aquatic microbe, *Loxophyllum*, is a thin, transparent protist which grazes on bacteria. It is one of the free-living organisms that help purify water. From time to time two cells meet, inspect each other and, if they approve of what they sense, they fuse together in an act of sexual reproduction. From the resulting zygote a new colony of the organisms is produced.

structure. *Loxophyllum* knows where it is going, and once in a while it will
cease to feed and seek out another *Loxophyllum*. Then the two will come
together on the oral side of the cells, joining together as a pair, and they will
then fuse and exchange nuclei in a microbial embrace. Should we conclude
that they enjoy sexual union? It is impossible that microbes do not enjoy
sex. If they did not, then they wouldn't indulge. Consider: they are brows-
ing around, feeding. They select what they wish to consume, and the more
food they eat the faster they keep growing and reproducing. To stop feeding
and search out a mate is an interruption of that vital process. So, at a cellular level, organisms like these ciliated protists must 'prefer' a little courtship
to mere feeding – something that we find reflected, I think, in our own
human behaviour.

The recognition of self and non-self and the whole-hearted extirpation
of non-self is so fundamental at the cellular level that this response
resonates in the behaviour of the organisms which are, in turn, composed
of such cells. It follows that this could explain exactly why groups of
humans feel compelled to attack each other. It is noteworthy that the most
anguished and appalling human strife takes place between very similar
groups of persons. During the Northern Ireland conflict we have seen
Christian pitting a pitiless war against Christian. They may be Catholic and
Protestant, but if you had some unbiased observer from another world,
they would go home after a period in Northern Ireland and report
that both sides of the conflict believe in the same religion. On further
questioning, your observer would insist that you could hardly find a single
difference – except that one group owe their allegiance to a cleric in Rome
and the others to someone in Canterbury. However, the religious obser-
vances were so similar, the rituals so close, that it would seem impossible
that so slight a change could make any difference at all. None the less, the
detection of differences within that community has led to vicious attempts
at mutual destruction. In Lebanon we saw the Shi'ite pitted against the
Sunni: both Moslem. The Christian and Muslim communities had always
coexisted amicably, and to mutual benefit. Suddenly, when territorially
dispossessed Palestinians were given sanctuary in Lebanon, the mutual
acceptance and respect disappeared, and factional fighting broke out.
Similar communities in Morocco coexist contentedly, and these cultural
enclaves had for long coexisted in Lebanon. A revealing example of conflict
arising through religious traits is the Gilbert and Ellice Islands of the
Pacific. Until they were colonized by Europeans, the inhabitants of this

great archipelago coexisted relatively peacefully. The European explorers introduced the religious split between Methodism and Roman Catholicism, since when the Gilbert islanders and the Ellice islanders have felt themselves to be in opposing camps.

The greatest benefit of modelling our behaviour through the interaction of single cells is that it explains where this appalling hatred has its roots. The most heartless acts of civil strife have nothing to do with territorial aggrandisement. There was an outbreak of violence in Los Angeles after the Rodney King trial, for instance, a perfect case study of what I see as cross-cultural conflict. No-one wanted to invade anyone else's territory. They wanted to express their hatred of the other cultural group – and hatred is the name of the phenomenon. In the bombing of the Sarajevo cemetery, nobody wanted to occupy the land but simply to hit so cruelly at people who are different. When the massacre at Kosovo took place, what we saw was an instinctual hatred of the Albanian culture by their neighbours. There are two populations that claim the name of Macedonia, and you might imagine they would like to combine – but no, as one has allegiance to Greece and the other claims autonomy, there is bitter feuding between the two. In Europe we have two major ethnic groups who are denied a home-land that they can call their own: the Basques, split between Spain and France, and the Kurds, straddling Turkey and Iraq. Both feel alienated and oppressed; the nations where they live express instinctual rejection of these people, and could not imagine creating a separate homeland. As a consequence, we are faced with the phenomenon of ethnic cleansing, a repugnant term which seems to sanctify the wish of one human commun-ity to extirpate another. Ethnic cleansing comes close to genocide, and we must get to grips with the motivations that have driven people to such terrible acts of inhumanity. We can now see that these are deep-seated reactions which have their origins in the nature of our very cells. We must also grasp the fact that humans have the power to comprehend these mechanisms. For an intelligent species like ourselves, it is important to understand cultural distinctions and accept them. If we understand the root of these feelings of conflict, we can at last accept ways to banish them forever.

Does this mean that Europe, as an economic and political entity, can never exist? Of course we can become closer. I doubt whether the forcing upon us of a single currency makes much sense at present, for the checks and balances we use within our national economies are so very different.

We impose duty of pounds on a bottle of alcoholic drink whereas some neighbours impose duty in pence, which makes a single unit of currency hard to rationalize. Laws on supply and demand vary greatly from nation to nation, and these need harmonizing first. We should allow our fiscal systems slowly to become co-ordinated, then currencies would adopt a stable relationship and a single unit will naturally emerge.

Can we perhaps find a biological model to explain the insistence on a single unit? I think we can. The European currency unit was originally going to be called, as though it were an acronym, the 'ecu'. The idea was promoted by the French, which to anyone versed in the way organisms like to use external markers could have revealed a hidden agenda. It was surely obvious that, if the term were an acronym for the European Currency Unit, the initials would have been in a different order were the term truly French. We would have been faced with the UCE, or even the CUE, but certainly not the ECU.

There had been an earlier example of the French imposing a marker on a subject of international importance. When the British design of the Concorde was jointly developed with the French, the name was going to be Concord (spelt the English way). The French managed to insist that it be spelt with a final -e, standing for Europe, they said. It was nothing of the sort, but was actually the infliction on an English-speaking aviation industry of a term spelt thus only in French.

The real reason the French were touting the idea of the ecu stems from the same wish to implant tokens, like cultural markers, around the world. Although nothing has been said about it, the ecu was a traditional French coin. The real écu (one word, for this is no acronym) was a solid silver coin worth six pounds. It was produced during 1792–93, and measured 39 millimetres (1½ inches) across. The original meaning of écu was 'shield', but it soon came to be the French slang term for 'cash'. By slipping it into the language of Europe, making it look new and non-partisan, the French were scoring over old rivals. I once suggested that we could adopt a different name, one that every Englishman or -woman would hold dear. I thought that we could call it the Quintessentially Universal Indicator of Denomination. There must be an acronym there, somewhere.

The issue of a single currency is deep and emotive, and is often taken as a mark of personal pride and national sovereignty. In any practical sense the fewer currencies the better, and we would be well advised to aim at global money and have done with it. It could save time and hassle, slash bureaucracy, reduce mistakes, cut the fiddles, and (best of all) eliminate the gross and mindless acquisitiveness of the speculators who dominate the world's markets.

In practice, there are drawbacks to this idea. In Europe, unlike America, commercial traditions vary greatly from state to state. There are nations, like Britain, where it is normal to own a home, and other nations such as France and Germany, where renting is more common. Some nations (like Germany) pay high pensions to retired persons; others (like Britain) offer far less. A single unit of currency, controlled by an agreed international interest rate, will have very different effects on nations with different fiscal and investment cultures.

The strength of a currency also reflects the perceived status of political power. The name it bears is a token of sovereign status, while the fiddles and loopholes people exploit in the switch from one currency to another is a sizeable tranche of the market. As the new unit of currency, thankfully renamed the Euro, will effectively have its base in Germany, it should be seen as a marker for that nation's culture. The aim of Germany was once the use of military might as a means of subjugating Europe and introducing economic control. In these more enlightened times, the conflict has been resolved in just this way, fortunately without further military conflict.

This nationalism and self-aggrandisement of the French surfaced again when the Euro was launched in 1998. According to the Maastricht Treaty, a head of the European bank must be appointed for a period of eight years. The unanimous choice of every European nation – apart from France – was Wim Duisenberg, the former head of the Netherlands National Bank. The French, who have felt under attack since their language fell from grace as the international language of academia and business, held out for their own compatriot Jean-Claude Tichet, head of the French National Bank. There seemed to be an impasse, because Tichet's appointment would not be approved by the other members of the European Community. In the event, the solution was that Duisenberg would choose to stand down after the first four years, and Tichet would take over for the last four. Although this seems to defeat the Maastricht Treaty, it was pointed out that the arrangement was voluntary, so the legal requirements of the Treaty remained (theoretically) inviolate. Like the infliction of the French spelling of Concorde, and the story of the écu, this forced solution by a single nation flying in the face of Treaty law and a unanimous block vote was a gesture. It is as though an external French marker is once again applied to an ostensibly neutral subject. Like external markers on living cells, or cats spraying a marker scent around new territory, the French feel that they have made their mark, and that is what counts.

The genes within cells can code for markers which allow them to recognize each other, and tell 'self' from 'non-self'. When the codes of recognition go wrong, we encounter autoimmune diseases. If cells of the body fail to recognize each other, and act as though self were non-self, fatal diseases can ensue. The most terrible of human diseases become easier to understand when they are seen as the reflection of the way that those single cells behave. Truly, we are nothing more vast than colonies of single and independent living things. The eternal warring within human society has an uncanny resemblance to what goes on between the cells inside our own bodies. The maintenance of our health depends upon the immune system. From the simplest forms of life upwards, single cells have the innate ability to recognize self and non-self. Non-self is rejected. If you suffer a splinter in your hand, then white cells will surround it and in time it will go away. But if you experience not a splinter but a kidney graft, the whole of your body's immune response will turn on that alien kidney, and destroy it. The recognition of self and non-self is one of the most fundamental propensities of single cells, and I believe that it is manifested in the way we interact. Genes create markers of identity on our cells, just as dogs like to detect each other by odour and create markers of their own as a means of identity. Producing a currency, or implanting a national word in another region, is displaying a mark of identity in exactly the same way. The study of the way cells interact will reveal the source of power behind sociobiology. At the very time when we believe that science has discovered almost everything worth discovering, I believe that we are just at the dawn of new science, a revival of real science: science that has descended from the realms of natural philosophy, and which fits together the massive amounts of data that surround us. Science in the decades ahead is going to be far more enlightening and enlivening than anything we saw from the years of the renaissance at its height. It may usefully remind us of the nature of human civilization and of the roots of the humility with which me must look at our fellows, and the self-criticism we might find in our hearts.

There is nothing inevitable about conflict in human society, if we use enough humanity to celebrate our cultural differences. Take an area like

Mediterranean Morocco. There you find orthodox and liberal Jews, Christians of all shades from Maronite to Catholic, and the Arab peoples, Tuareg and Bedouin, all coexisting as neighbours. The USA celebrates the cultural disparity between her peoples, and nations like Mauritius are composed of different cultures that are accepted for what they are, and which harmonize in daily life. I believe that our only way to progress in the future is to abandon the foolish belief that everyone is the same. This is dangerous and impracticable. Blacks really are very different to whites. Women really are very different to men. The Welsh are culturally quite disparate from the English. The Germans really are typically Teutonic just as the French really are somewhat arrogant and self-seeking. The Spanish are a hedonistic race who are disinclined to punctuality, just as the Italians are amorous and extroverted, the Americans friendly and parochial, and we English are the most hypocritical race I have ever met in the whole of my travels. I am not merely noting the problems with the others – I am just as glad to celebrate the adverse aspects of our own culture. In countries where the differences between cultures are noted, their benefits valued and respected, and where people know that they are distinct, then the problems begin to go away. The problems arise in nations where you have different cultural groups, but nobody will admit it. These people are all forced in to the same mould. That is why we have had difficulties in Lebanon and Northern Ireland; this was where people were not encouraged to celebrate each other's own cultural distinctness.

We have observed the acquisition of cultural markers within recent decades. When Germany was split in two there was an international sense of outrage. The blockade of West Berlin (which became an enclave within the German Democratic Republic) led to a continual airlift of vital supplies. Families, divided by a wall, waved piteously at each other. Neighbours were split apart by concrete and barbed wire. On my visits to East Germany, the greatest sense that prevailed was one of lost contact with loved ones in the West. Then the wall came down. As we stood under the Brandenburg Gate on *Enheitsnacht* (unity night) in 1989, flags waved, fireworks blasted the night sky at midnight, armed troops melted into the shadows as the two-million crowd remained happy and laughing. The future looked promising. Since then, the effects of the period of separation have become apparent. There are wage differentials, imposed by Bonn, against those who lived in the East. The East Germans (the so-called Ossies) look askance at the brash materialism of the Wessies. The Wessies look upon the Ossies as inferior beings. The simple act of holding the community in two halves for a generation gave rise to enough cultural distinction to create new enemies of old friends.

It is our differences that make us what we are. Yet still we believe that it is some idealized 'sameness' that underpins human society. Some of the lyrics from popular music recapitulate this propaganda. The band Blue Mink, in their recording of 'Melting Pot', extolled the inevitability of 'coffee-coloured people by the score'. We had 'Ebony and Ivory' by Paul McCartney and Stevie Wonder (the singers 13.000 km (8,000 miles) apart, because they never met when the tracks were laid down) emphasizing that, beneath the covering of skin, we were all exactly the same.

At the end of the line, we are not all culturally coffee coloured. Racism is a terrible scourge, and we need to educate people to its realities. In distinguishing one race from another we must never make the age-old mistake of believing that one race is superior to any other. The races of humankind are not identical, though they are equivalent. Celebrating the differences must never mean inculcating a sense of superiority, for therein lies the seed of Nazism. With this understanding at our side, nobody can retreat into believing that their culture is somehow the best. Commentators cannot find refuge in a belief that racial hatred expresses low self-esteem on the part of the aggressor. A rejection of non-self is an inborn part of life, but as humans we have the ability to embrace humanity, trust and understanding. It is in the ways of our single cells that we may perceive ourselves. The culture of micro-organisms can yet reveal new aspects of the cultures of humankind.

10

Hopes for the Future

WE FACE AN EXHILARATING FUTURE. Now all we have to do is relate the concept of the single cell to our understanding of human behaviour. Cells have grouped together to enable them to fill new ecological niches. Develop cellulose cell walls and you can become a plant and stand tall. Form parts of yourself into fins and you can swim; develop some into lungs and you may walk on dry land. Within our bodies lies that inner salt sea in which early life existed, and the purpose of sense organs is to maintain that system. Partly, this is done by finding a mate and identifying them, but on a daily basis our senses find expression through locating a supply of food and water. This is the purpose of a human body, which exists primarily to propagate germ cells that bequeath something of our nature to the next generation. This is not the only purpose of living, because the childless, the committed homosexual and the bereaved play parts of equal societal value. In some cases, the creative impulse that most people express through new generations of children finds an outlet through other manifestations.

Although we can be seen as mere expendable fruiting bodies, the true *Homo sapiens* is immortal for it never dies. This is true of the ovum and the spermatozoon. Let us consider the sperm, if only because the ovum does nothing but lie there, whereas the sperm actually does things and is off exploring new areas of territory. The ovum is a highly complex resting cell. If you were to isolate a single live spermatozoon, on a microscope slide, and someone were to ask its identity of a microbiologist – who, on this one occasion, had never seen a sperm before – they would say it was a flagellated microbe: a protist of some sort. You might persist and ask of what sort. And your microbiologist would tell you that there were thousands of them in pond water, and it could be almost any one of countless groups. It is no more than a microbe with a flagellum, swimming about. Yet that is the cell that holds the essence of humankind. It is the same with the ovum. The naïve microbiologist confronted with this cell would conclude that it must

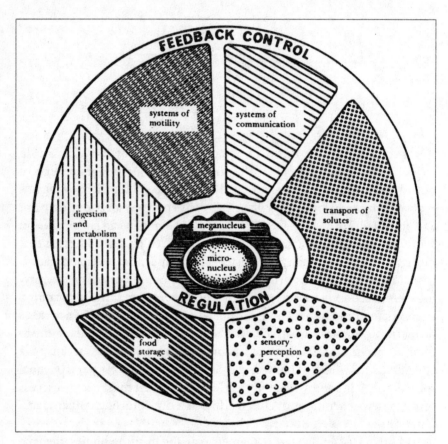

The division of labour within a single cell. One way to compare a protist with a multicellular organism is to construct a diagram showing how the different functions of life are contained within the cell. Different parts of the cell can carry out movement, sensation, digestion and food storage, for example. There are complex feedback mechanisms that keep the whole system in equilibrium. The genes are inside the central nucleus. In some protists there is a micronucleus (for the genes) and a meganucleus to control the day-to-day running of the cell.

be a resting stage of an organism because it was not actually doing anything. It has no cilia, it doesn't crawl, it just lies there rich in food reserves. But, it too, could be a single-celled microbe, a protist.

There is nothing about the appearance of these cells to suggest that the two might fuse and that this innocent globular cell might become infected by a swimming germ which comes along and causes the hapless ovum to enter a phase known as embryological development. Nobody would guess that these ordinary-looking single-celled entities would fuse and give rise to people. The only reason that they do give rise to people is because we are the fruiting bodies on whom the survival of these simple protists depends. These organisms, swimming in their saline environment of ejaculate, are swimming around in water as they have always done. Indeed ejaculate, like tears, is slightly salty and this is because these life forms evolved in the sea. They always were used to living in the sea, and so they have developed a form of larger fruiting body that can generate and enclose a portion of sea in which they can survive. That spermatozoon is the microbe *Homo sapiens*, swimming in the sea as it did a thousand million and more years ago. The moribund fruiting bodies are the expendable husks which are supervened by the immortal sperm. That is the true nature of our kind.

I have referred to the difficulty philosophers have had in relating a zygote to an adult. On the one hand, the fertilized egg cell is simple and uncomplicated, because it is a single cell. Yet it contains all the genetic information needed to build Concordes and fly in them. How can the two – the single cell and the multicellular adult – be related? The philosophy of this question has dogged science for a very long time. Here is an answer. Consider a single protozoan cell. We may view it as a schematic system of separate functions. This model does not fit well with an ovum, because that cell is in a resting stage and waiting for events to start. But if you imagine a spheroidal typical cell, then in the centre lies the micronucleus (we discussed this in the case of *Loxophyllum*), the small nucleus that contains the essential DNA. Around it lies the meganucleus which regulates what the micronucleus does to translate its actions into the propensities that give rise to the phenomenon of a complicated and mature cell.

Spread around this core we would have specialized regions providing the following functions:

* Systems of communication, because protists need to communicate.
* Transport of solutes, because cells (once they have taken in a food

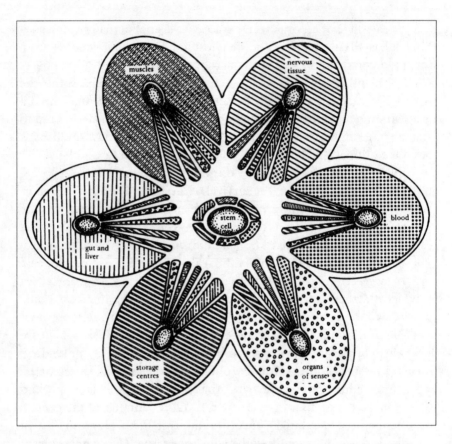

Relating the single cell to an organism with millions. In a many-celled organism we can show how each cell could develop a specific ability that is appropriate to its role in the body. The motility is concentrated in muscle cells, sensation in sensory neurons, etc. This unifying model also allows us to deduce that the properties of single cells are reflected in the behaviour of a multicellular organism.

particle) need to digest it and then to translocate the digested components to different parts of the cell.

- Sensory perception, because protists need to sense their surroundings – they will detect and swim agitatedly away from unpleasant stimuli, which shows that they have recognized that the stimulus was unpleasant.

- Food storage, of which the ovum has an abundance whereas the sperm has almost none. Protozoan cells have a centre for food storage because you never know when you might need it. The food reserve bodies within the protists are the beer-gut equivalent of the middle-aged human male.

- Centres for digestion and metabolism, which are self-explanatory.
- Systems of motility, which in the ovum are not developed and which most resting cells similarly lack. But many animal cells can move, and we are made of cells ourselves. If you look at time-lapse film taken of human cells grown in vitro on television, you will see them moving about in a state of great activity. They are often on TV, though narrators rarely tell you that they are time-lapse films, so it is easy to be given the impression that this is what cells in real time are really like. It should be emphasized that the movements are normally slow, and only look so frantic when considerably speeded up.

So, how can we relate these properties of a single cell to an adult, many-celled human being? The way it works, according to this model, is that, in many-celled organisms, the specific functions within a single cell are delegated out to cells that become primarily devoted to a specific task. Let us consider the systems for motility. The motile proclivities within the single cell are translated into the specialized functions that take over the striated muscle cells. This is what a muscle cell does above all else, and it retains only a limited ability to carry out any other functions. Neurons have developed their 'neuronic protein' until the cells have become almost entirely devoted to carrying out the transmission, analysis and processing of information on a vast scale that goes on within each cell. Similarly, the transport of solutes within the single cell is the function specialized in the blood system. In this way we can explain the relationship between the ovum and the adult that has puzzled biologists for centuries, ever since a description of the ovum was first published in 1672 by Leeuwenhoek's mentor Reinier de Graaf (1641–73). We must abandon the idea that the cell in some way forms a body that becomes a person; instead, this person results from the behavioural manifestation of the separate cells of which we are all composed. The microscope substantiates this view: the cells from which specialized cells develop (the stem cells) are similar to embryonic, unspecialized cells. It has been traditional to regard these cells as unspecialized, but it may be more sensible to conclude that they are actually pluripotential. It is not that they cannot do much, but that they can do a little of everything.

Within the adult body, it is important to realize that cells make their own decisions, and act without control by the whole of the body. If you have a tooth extracted, you may find a small portion of the bony socket is left projecting above the gum. Sometimes a fragment of the root breaks off and is lost within the empty socket. With the classical teachings of the way a whole body responds, you would expect the projecting bone to remain as a hazard; a piece of root detached within the socket could act as a focus of infection for years ahead. This is compatible with current medical understanding, but, if we look at these events from the viewpoint of the neighbourhood of the single cells, the fate of each condition becomes very different. The cells cluster around a portion of projecting bone, and systematically decalcify it. At the base of the projection, osteoclasts congregate. These cells break down bone and absorb it. Within a week, the bone has gone. The cells of the gum proliferate in a co-ordinated fashion and cloak the site in new tissue. Within a month you'd never know that the damage had occurred. The fate of a broken piece of root is similar. Within a few weeks of the extraction, the fragment emerges at the surface of the healed socket and the gum closes seamlessly behind it. Cut a branch from a holly tree and something similar happens. Cells from the surrounding cambium are brought to full activity and slowly close the gap. Within a year or two the cut is healed, and the damage has been controlled. To pretend that these systems have somehow evolved makes no sense at all. Teeth are not extracted in nature, any more than branches are sawn cleanly away. What is happening is that the cells in these regions are self-motivated and act in a co-ordinated fashion to restore the body to soundness. None of this is under any nervous or hormonal control, and these processes go on in the private world of the cells, without reference to the organism as a whole.

Communication between individual cells has now been shown to be important in the healthy eye. In some cases of corneal opacity the cells within the cornea gradually stop communicating with one another. This single lack of communication disrupts the local environment within which the cells exist, and the imbalance causes cloudiness and eventual problems with eyesight.

Cells also know when to die, and programmed cell suicide is an important part of life. We saw how the slime moulds such as *Dictyostelium* move together to form a single organism and then produce an aerial stalk, a sporangium, and a great cloud of spores that can be released into the wind. The cells learn what role they are expected to play, and those that form the aerial branch are sacrificed in the cause of reproduction. Within our own bodies, similar processes take place. We are born with a gland in the chest,

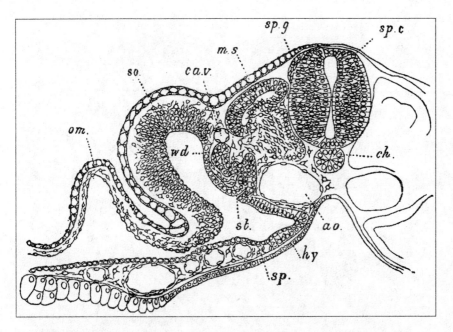

Cells in the early embryo take up position for adult life. The beautiful choreography of a developing embryo was first studied over a century ago. Here we see cells in a developing duck embryo. This sectional view was published in the *Quarterly Journal of Microscopical Science* in 1875. You can clearly see the well-developed spinal chord ('sp.c') and the start of the backbone ('ch'), the aorta ('ao') – the body's principal blood vessel – and early bands of muscle ('m.s').

the thymus, which slowly disappears during adulthood. It lies beneath the breastbone, and is shaped like a leaf of thyme (which is why it is called the thymus). This gland is an intensive centre for producing and training lymphocytes to recognize friend from foe. The lymphocytes that derive from the thymus are known as T cells (T for thymus) and they are key fighters against disease. Interestingly, many of them remain within the gland rather than circulating around the bloodstream: they live and die within the cortex of the gland. We know how important they are because, if the thymus is missing during childhood, a slow wasting disease occurs. In adults, however, the gland is much smaller and fat replaces the spaces where the lymphocytes were once found. This slow loss of the thymus is also due to programmed cell death. Many scientists believe that this phenomenon is as important as cell division. The scientific term for programmed cell suicide is apoptosis. It is pronounced without the

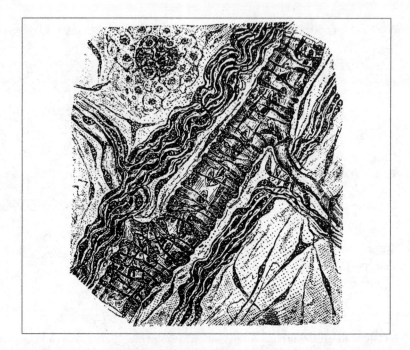

Specialization of developing animal cells. This vivid illustration of living cells from a Victorian textbook shows clearly how genes signal to cells as they specialize. This is the transparent web of a frog's foot, which can be examined easily without harm to the frog. Lionel Beale, who made these studies in the 1850s, described the blood vessel (centre) and the bundles of nerve fibres on either side. To the lower right are separate fibrocytes, the cells that help hold the tissues of a mature animal together.

second 'p' being sounded, as 'appo-tosis', and they say that mitosis and apoptosis need to be in perfect balance for the maintenance of a healthy adult body.

One of the most important things for a cell to know is when not to divide. Tumours form if cells multiply when they shouldn't. Should cells spread through the body from the edge of the tumours, we end up with cancer. In some forms of the disease a gene has been found to be responsible. In others we know that the cancer is caused by a virus. Leukaemia in birds is caused by viruses, for instance, though we do not find these viruses in human leukaemias. Perhaps we will, in time. When the first cancer genes were discovered they were named oncogenes (oncologists are the specialists who study cancer). After an initial rush of enthusiasm, it began to look as though oncogenes were involved with only a few forms of human cancer. We have tended to look at cancers as though their cells have received a boost, and have super-powered growth genes. The cancer cell, in this way, is like a car when the accelerator pedal is used to increase the speed.

I used to prefer a different way of looking at it. Imagine the cancer cell as a car whose engine is always trying to drive it on. Its speed is not controlled by an accelerator, but by the brake. Normal cells apply the brake at the right time. In this case, a cancer cell could result not from too much gas, but from not enough braking. We know that many of the substances that cause cells to become cancerous are compounds that damage cells. It can be hard to imagine how something that damages a mechanism could make the mechanism work faster. If the natural behaviour of a system is to travel fast anyway, then it is much easier to see how damage to the brake could cause it to travel on out of control. There would also be the possibility of curing

Does a three-year-child old draw a protist cell? The drawings of little children sometimes seem to evoke images of single cells. We all began as single cells, and embody resonances of their behaviour throughout our lives. Perhaps the seemingly random drawings of the untrained mind reveal latent memories of our cellular inheritance.

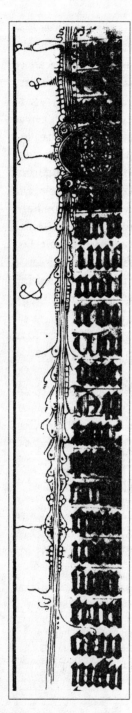

the cancer. In the 'accelerator' model we have to cut off the fuel supply. In the 'brake' theory what we have to do is fix the brake, and bring the cell back under control. Studying single cells and how their genes can be turned on and off brings us closer to understanding cancer.

Throughout our lives, we are more closely connected with the single-celled microbe world than we realize. Not only do we become microbes when we undergo reproduction, the most crucial episode in our lives, but our genes may confer upon us instinctive intimations of the sense of awareness of single cells. In the world of art, there are pictures that embody organic shapes, some of them seeming to embody the structure of cells in the mind of the artist. Although it seems sensible to assume that these conventions have been learnt by an acquaintance with popular science, there are other examples that are less easily dismissed. Children sometimes create drawings which look like images of cells. During my recent studies of mediaeval manuscripts, I was drawn to study the decoration around the edges of an illuminated text. The pictures were cellular. They look like tissues, at the cell's eye view. Here was an image that seems to be drawn from an awareness of

Mediaeval embellishments and cell memory. Many early documents were elaborate, decorated with marginal illuminations. In some cases these free-wheeling embellishments may embody subconscious resonances of our life as single-celled organisms. This astonishing mediaeval manuscript, from the library of Skip Palenik of Illinois, has marginal illuminations which portray cell-like images. Flagella and nuclei seem unmistakable. The capital letter E, for example, shows something like a sperm fertilizing an ovum. Our cellular nature may be echoed in these complex drawings.

what a living cell is like, and yet was made several centuries too soon to have derived from a microscopic study. Are there resonances here, too? There is a classic psychological picture test that shows a dog wagging a black and tapering Mickey Mouse tail which is threatened by amputation with an axe poised above it. The image is said to be threatening, and men are said to recoil from the image instinctively. It is concluded that this is fear of castration. It may be nothing of the sort. It is only a picture of a dog about to have its tail chopped off … but the tail is long and thin. It is just like the tail of a sperm. Could that be the echo in the mind? There are some animals that seem to make most women react: mice and snakes. The tail of a mouse, and the sinuous nature of the snake's body, are both reminiscent of an approaching spermatozoon. There could be a resonance between the way in which the sperm enters the ovum during fertilization, and the entering of the woman by her male lover. Psychological explanations constructed upon these grounds may make more sense than some classical conventions. Memories of pre-fetal existence are claimed by some people, and for all its improbability there may be a sliver of sense in all this.

Does this new concept mean that warfare is genetically programmed? This is seen by some people as a justification of war. Unesco published a statement on race, which denounces any justification for racist behaviour on the grounds of it being genetically inborn. The most recent expression of this view was the Seville Statement of 1996, laying down that it is scientifically incorrect to say that human beings are genetically programmed to make war. The truth of the matter is that the delegates at Seville did not really know whether the genetic programming exists or not. There is no research into the matter, and no likelihood of genetic research that could make any sense. In reality, official bodies prefer to adopt an unjustified stance of 'science' when the real science does not exist. Humans know that they are able to benefit their fellows, to protect the planet, to nurture the natural world; acts of war are conscious decisions over which reason can mediate.

Humans are cognitive, sophisticated, intelligent, intellectualizing organisms. You may try to claim that any form of inhumanity is hidden in your genes, but that does not justify it. It is true the Y chromosome codes for testosterone, and that testosterone is a hormone linked to aggressive behaviour, but warfare is a corporate enterprise. Those who conduct wars are remote from the soldiers whose lives are prejudiced. There are no genes to make us erect factories that build munitions, and make huge profits for the war lords. However, I have here presented evidence from many sources that rejection of non-self – and a deep-seated wish strongly to reject the very similar non-self – is an inherent component of the nature of life: an engine of biological development, the secret behind the flourishing of languages and the key to speciation. This is a biological imperative that exists at every level of life. It helps to explain the origination of our impulses, but this cannot justify our acting towards each other with inhumanity. We are supposed to learn social interaction, and civil strife occurs when this vital part of our learning is undone. Here, too, it is intriguing to examine other forms of life, and to discern parallels between how humans behave and the way our cells sometimes seem to act.

In recent years, in human medicine, much attention has been focused on the study of the autoimmune diseases, when the cells within the body start to attack other cells with whom they ought, in fact, to coexist. There are sadly many recalcitrant diseases that come into this category. Attention is being paid to these syndromes of self-destruction because of the need to control the rejection of non-self in transplant recipients, and to solve the tragic burden of autoimmune diseases. I see a close analogy between the way that cells interact antagonistically, and the way in which the bodies of cells that grace this earth also do so. Dogs fight much more violently together than a dog will fight with a member of another species; indeed, a dog fight is the most violent interaction in which you would find your pet engaged. Dogs kill prey, yes; but dogs fight because other dogs are similar in being dogs, yet being different in appearance and in character. I believe that, even at this elementary level, slight differences can be seen as exerting the greatest effect when we consider how slight non-self distinctions are mirrored by the microbial world. The internecine societal strife we see around the world is an autoimmune disease – this time of people. The mechanisms in our genes are intended to protect us, but were not developed under the pressures of modern society.

The results of the identity with self – and the rejection of non-self – is daily evident around us. During 1998 we saw the return to Britain of several persons convicted of killing in foreign countries (in one case the victim was a nurse, in another a baby). Widespread unease surrounded both cases. Few people seemed to dispute the gravity of the convictions. None the less, the returning women were hailed almost as rock stars or media personalities. The reason is because they were convicted by foreign courts and the crimes were committed in alien cultures. Had they been centred on home territory our feelings would have been very different, but because the convicted persons were 'our' people tried in 'their' courts they received widespread media support.

Yet we have also seen that altruism exists in the genes. The great cohort of separate cells that combine to raise the slime mould sporangium above the forest floor shows that self-sacrifice is inherent in the community of cells. The phenomenon of apoptosis shows that cell suicide and programmed cell death is a vital aspect of the life of multicellular organisms, and the way that parents risk their lives to save their young shows how altruism is part and parcel of life. We do not need to cite examples, because reports of dogs, cats and even birds show that the parents jeopardize their lives if their offspring are at risk in a fire or flood. The whole of nature features examples where disparate organisms act for mutual benefit, rather than mere exploitation. Bacteria convey health on multicellular organisms, fix nitrogen for vascular plants and interact with other organisms to fertilize the soil. Fungi and algae co-operate to form lichens which alone can colonize rocky outcrops and bare stonework.

We do not have to look far to find altruism in human society. Carers care for their charges with devotion, and often far beyond the attraction of monetary reward. There a 'selfish' explanation for all this, of course, which has been mirrored by the idea of the 'selfish' gene. People do things to please themselves: if it gratifies someone to be a carer, then their apparent altruism is in fact an act of selfish gratification. This explanation belongs in the same category as 'all life is a dream; I am dreaming this now'. You could argue this way, but there is no reason to invoke such surreal explanations in the face of the devotion that people show to their fellows. Nurses and paramedics (such as physiotherapists) often work far above the demands of the job to help patients and to ease the burden of suffering. Did that brave man who leapt into the icy Potomac river to rescue a woman from drowning do it to get on TV? He was risking his genes in the process. Was he perhaps being 'driven' to protecting hers? For her genes to urge him towards self-destruction would be a magical transfer of the 'selfish' impulse.

The caring professions, often relatively lowly paid, show us human altruism at its best. Could it be that such people are doing it to earn the respect of their subjects? Animal welfare provides a further timely reminder of the selfless devotion of individual people. The determinists will perhaps say that they are seen to be acting thus and earn respect, so

1	2	3	4	5	6	7	8	9	10
0 sr (striate leaves)	0 ws_3 (white sheath)	0 cr_1 (crinkly leaves)	0 de_{16} (defec-tive seed)	0 a_2 (antho-cyanin)	0 po (poly-mitotic)	0 in (inten-sifier)	0 v_{16} (vires-cent)	0 Dt (dot-ted)	0 Rp (resist-ance to Puc-cinia)
15 ga_2 (gameto-phyte factor)	11 lg_1 (ligule-less)	18 d_1 (dwarf plant) centromere		6 bm_1 (brown midrib) centromere	13 Y_1 (yellow endo-sperm)	4 vs (vires-cent)	14 ms_8 (male sterile)	7 yg_2 (yellow green)	16 Og (old gold stripe)
25 ms_{17} (male sterile)			35 Ga_1 (game-tophyte factor)	7 bt_1 (brittle endo-sperm)		18 ra_1 (ramosa ear)			
27 ts_2 (tassel seed)	30 gl_2 (glossy leaves)	32 rt (root-less)		10 v_2 (vires-cent)	33 pg_{11} (pale green)	22 gl_1 (glossy leaves)	28 j_1 (japon-ica stripe)	26 C (aleu-rone color)	28 li (lineate stripe)
28 P (peri-carp color)		40 Rg (rag-ged leaves)		12 br (brevis plant)	44 Pl (antho-cyanin)	32 Tp (teopod)		29 sh_1 (shrun-ken endo-sperm)	
30 zl (zygo-tic lethal)	49 B (antho-cyanin booster)	47 ts_4 (tassel seed)	56 Ta_1 (tassel seed)		45 Bh (blotch-ed aleu-rone)	36 sl (slashed leaves)		31 bz (bronze antho-cyanin color)	38 ls (luteus)
53 as (asyna-psis)	56 sk (silk-less)		66 sp_1 (small pollen)	31 pr (red aleu-rone)		38 ij (iojap stripe)			43 gl (goldeu plant)
59 hm (helmin-thospor-ium re-sistance)	68 f_1 (floury endo-sperm)	64 ba_1 (barren stalk)	71 su_1 (sugary endo-sperm)	40 ys_1 (yellow stripe)	54 sm (salmon silk color)			44 bp (brown peri-carp)	57 R (antho-cyanin)
75 f_1 (fine striped leaves)	74 ts_1 (tassel seed)	75 na_1 (nana plant) centromere	74 de_{16} (defec-tive endo-sperm)		64 py (pigmy plant)	56 Bn (brown aleu-rone)			
79 Vg (vesti-gial glumes)			84 zb_6 (zebra stripe)	72 v_1 (vires-cent)				59 wx (waxy endo-sperm)	
80 br (brach-ytic plant)	83 v_4 (vires-cent)	103 a_1 (antho-cyanin)	100 Tu (tuni-cate)						
102 an_1 (anther ear)		115 et (etch-ed aleu-rone)	103 j_2 (japon-ica stripe)			92 bd (branch-ed silk-lees)		71 v_1 (vires-cent)	
129 gs_1 (green striped leaves)	128 Ch (choco-late peri-carp)	121 ga_7 (gameto-phyte factor)	111 gl_4 (glossy leaves)						
156 bm_2 (brown midrib)									

self-interest still prevails. Yet people sacrifice their lives to protect those of animals. There is little self-interest in a heroic death. We see acts of genuine devotion, and altruistic behaviour, all around us every day. The notion that altruism is really just selfishness is one of the reasons why people distrust each other these days. We are mutually supportive and helpful beings. Co-operation and interdependence are part of our inheritance.

Just as people should positively interact, the cells within our bodies act in concert during our lives. They have to process information and know what to do next, as we go on conquering new lands and flying through space, inventing new machines to help society, and trying to make sense of the world. The well-ordered co-operation within our own cell communities is what makes us human. Our behaviour reflects their behaviour. Once we understand how cells behave, we will discover our own behavioural roots.

We must not be frightened by the concept of genetic engineering, but should learn the importance of what it could do. The argument that it is a matter only for God, and not for humankind, is hard to sustain. Religious faith must surely accept that human intelligence is the highest manifestation of a creator's purpose, and the production of new species – from dogs and cats to wheat and maize – is among our predecessors' highest achievements. God has given us the ability to investigate the gene, yet we must ensure that crude commercial interests do not misuse this for short-term expediency. In servicing shareholders, we must never neglect our duties to the future.

We have already addressed the suffering of childlessness, and we embrace the use of modern technology to terminate pregnancies at will and to start them where otherwise they might not be able to occur. We artificially transplant children from one family into another by means of adoption, itself a highly unnatural procedure, and rely on species that have been created by our forebears for our survival.

Opposite: **Important genes on the chromosomes of maize.** More is known about maize genes than about those of almost any other crop. This gene map of ten maize chromosomes dates from 1935, by which time M.M. Rhoades had already mapped hundreds of important genes. Modern methods allow us to position new genes accurately on the chromosomes, and to change those we do not like. However, commercial interests should not be encouraged to impose new conventions upon world agriculture.

To live in something as sophisticated as a civilized society compels us to use highly unnatural processes. Genetic engineering offers an end to much human suffering and will convey valuable insights if we use it well. Changing a gene will make it possible to cure many of the most dreadful diseases that afflict the suffering. We will find genes related to artistic abilities or to memory, and to the cause and cure of cancer. It is a brilliant and exciting prospect. We have seen that altruism is a fundamental attribute of life, and these powerful discoveries cannot be bought and sold in the spirit of Victorian trade. However, we must seek to avoid the misuse of science to make money for speculators to whom short-term profit matters more than long-term risk. The question of genetic prejudice will also arise. Under current conventions, there would be nothing to prevent life assurance companies from asking about genetic susceptibility and offering a lower premium to individuals free of specified disease-inducing genes. Do we really want a future society where genetics dictate personal value? Can we allow commerce to decide the future of genes, imposing world-wide monopolies – as could easily happen in agriculture?

There is nothing inherently fearful about genetic engineering, any more than there is intrinsic danger from television or cars, medicine or technology. In each case we have structures set out to prevent the misuse of anything new, for it is abuse that poses the threat, and not the nature of the facility. Clearly, we need far-sighted, sensible and realistic regulations. We allow British people to smoke tobacco, but not marijuana; to buy alcohol, but not beef on the bone; to sell paracetamol (which can damage the liver) but not ecstasy (which can damage the brain and the bank balance). We already interfere in reproduction to an amazing extent, and now must recognize the reality that genetic engineering is here to stay. This new science marks out our future, and society needs to decide sensibly on its control.

Let me emphasize that many of the most controversial developments in the biosciences have nothing to do with genetic manipulation. The transplant of a new hand on to the amputated stump of 48-year-old Clint Hallam in September 1998 may offer hope to amputees. Although it owes much to the skill of doctors trying to suppress rejection of the transplant by Mr Hallam's immune system, it owes nothing to genetic engineers. Meanwhile, a team of French scientists is experimenting with flightless ladybirds to aid pest control. Their ladybirds were bred from wingless varieties which were hatched after bombardment by nuclear radiation, but that is just a matter of chance and no DNA transplanting was involved. In

Britain it is planned to produce hens' eggs low in cholesterol and high in the fatty acids we need to keep healthy. That is being done by modifying the diet of chickens, and not by tampering with their genes. There are commercial clinics that aim to implant embryos into would-be mothers, and even to try to decide the sex of the child. But there is not genetic engineering there, either.

In my view, we will learn most about our nature by looking at our cells as living entities with a mind of their own. It is single cells that govern the biosphere; the genes they use to pass information on to successive generations are their messengers, not their masters. I want us to celebrate our cellular immortality, so that we can make the best of our existence as expendable reproductive bodies whose sole function is to integrate with our fellows and improve life's lot. It is as a microscopic cell that we began, as microbes that we survive into new generations, and it is microbes that will recycle our remains for succeeding life forms. We are cosseted and fed by microbes, our hospitable environment is created by them, and our destiny is to become them once again. Every day of our lives is imbued with the resonances of how a single cell behaves. We are part of the web of life, and our genetic nature links us to every single living thing scattered across this wonderful and captivating planet.

Index